Geometrical
Properties of
Vectors and Covectors

an introductory survey of differentiable manifolds, tensors and forms

Geometrical Properties of Vectors and Covectors

an introductory survey of differentiable manifolds, tensors and forms

Joaquim M Domingos

University of Coimbra, Portugal

 **World Scientific**

NEW JERSEY · LONDON · SINGAPORE · BEIJING · SHANGHAI · HONG KONG · TAIPEI · CHENNAI

Published by

World Scientific Publishing Co. Pte. Ltd.

5 Toh Tuck Link, Singapore 596224

USA office: 27 Warren Street, Suite 401-402, Hackensack, NJ 07601

UK office: 57 Shelton Street, Covent Garden, London WC2H 9HE

British Library Cataloguing-in-Publication Data
A catalogue record for this book is available from the British Library.

GEOMETRICAL PROPERTIES OF VECTORS AND COVECTORS
An Introductory Survey of Differential Manifolds, Tensors and Forms

ISBN-13 978-981-270-044-5
ISBN-10 981-270-044-7

Printed in Singapore

Preface

From the beginning of the XXth century, physicists were faced with the outbreak of new languages in the mathematical description of physical space. This book is aimed to give a short and elementary contribution to elucidate the connection of some geometrical topics taken from Manifold Theory with the more familiar calculus on Euclidean space. The pretext is to make accessible the relationship of the wedge product of covariant vectors (1-forms) with the familiar cross product of contravariant vectors.

It is appropriate to say a word about the organization of the book.

Chapters 1, 2 and 3 contain some basic topics on topological spaces, metric structure and differentiable manifolds. The manifold is made up of patches by smoothly pasting together open subsets of a topological space which are homeomorphic to open subsets of $\mathbb{R}^n$. The notion of tangent vector to a differentiable manifold, at a point, is viewed as a directional derivative operator acting on functions. The existence of a moving frame on a manifold is discussed. Chapter 4 is mostly about the metric dual operation, induced by the metric, which establishes a 1-to-1 correspondence between vectors and 1-forms (covectors). Chapter 5 is concerned with the basic properties of tensors, particularly covariant tensors. In Chapter 6 r-forms, i.e., the antisymmetric covariant tensors, are treated in some detail. In Chapter 7 the property of orientability of manifolds is dealt with.

In Chapter 8 given a metric and an orientation, we introduce the Hodge star operator which defines a canonical isomorphism between r-forms and $(n-r)$-forms. Finally in Chapter 9 we clarify, in terms of the Hodge star operator compounded with the metric dual operator, the conditions under which the wedge product of covectors (1-forms) produces the cross product of vector algebra.

In the course of writing this work a great deal was owed to the books listed in the Bibliography.

Joaquim M. Domingos

Contents

Chapter 1

Topological Spaces

"Topology is qualitative mathematics"
M.J. Mansfield

Let X be an arbitrary set whose elements will be referred to as "points". If we wish to express the degree of nearness of these points, we must introduce a *metric* on X,

$$d : X \times X \to \mathbb{R}, \qquad (1.1)$$

assigning to each pair of points $x, y \in X$ a number $d(x, y)$. This number is called the *distance* between x and y, and any set X endowed with a metric is called a *metric space*. The number $d(x, y)$ has to satisfy, for all $x, y, z \in X$, the properties

- $d(x, y) \geq 0$; $d(x, y) = 0$ if and only if $x = y$
- $d(x, y) = d(y, x)$
- $d(x, z) \leq d(x, y) + d(y, z)$.

The differentiation of a function at a point x is a local operation, requiring values of the function in the neighborhood of the point at which the derivative is taken. To support concepts such as continuity and differentiability we need to impose a decomposition of the space into a collection of neighborhoods. A neighborhood of a point x of a metric space X is considered to contain all the points of X sufficiently close to x. The concept of open ball enables us to give a more precise

1

definition of neighborhood of a point. Let x be any point of a metric space X and let ϵ be a positive number. The subset $B(x, \epsilon) \subset X$, consisting of all points y in X whose distance from x is less than ϵ

$$B(x, \epsilon) = \{y \in X | d(x, y) < \epsilon\} \tag{1.2}$$

is called an *open ball* about x.

A subset (not necessarily open) of a metric space X is a *neighborhood* of a point x in X if it contains an open ball containing x. The subset U of a metric space X is *open* in X, if for every $x \in U$ there is $\epsilon(x) > 0$ such that the open ball $B(x, \epsilon)$ is entirely contained in U; that is, if U is a neighborhood of everyone of its points.

The metric space X is open in X, in view of the fact that we cannot go further away from it. That is, the whole metric space X is a neighborhood of each of its points. A point x is said to be an *interior point* of a subset A of X if A is a neighborhood of x. Therefore, A contains entirely an open ball about x. The *interior* $\mathring{A}$ of A is the set of all interior points of A and being the union of open sets, is an open set in X. Hence, the subset A is open if $A = \mathring{A}$. As the empty set $\emptyset$ has no interior point, $\mathring{\emptyset} = \emptyset$ and $\emptyset$ is open. So, for a metric space X, $\emptyset$ and X are both open in X.

Given two subsets $A \subset X$ and $B \subset X$ with $B \subset A$, the set

$$A - B = \{p \in X | p \in A \quad \text{and} \quad p \notin B\} \tag{1.3}$$

is called the complement of B in A. A set is said to be closed if its complement is open. As

$$X - X = \emptyset$$
$$X - \emptyset = X \tag{1.4}$$

the two sets $\emptyset$ and X, which are both open in the metric space X, have complements X and $\emptyset$ which are both closed. That is, X and $\emptyset$ are simultaneously open and closed.

A *mapping* of the set X into the set Y, $f : X \rightarrow Y$, is a rule which associates to each element of X an unique element of Y. Let A be a subset of Y. Then, the subset of X denoted by the symbol $f^{-1}(A) = \{x \in X | f(x) \in A\}$ is called the *inverse image* of A under f.

One might say that it is the union of the sets composed of all $x \in X$ for which $f(x) \in A$. We should notice that the symbol $f^{-1}(A)$ does not define a mapping of the subset A under f^{-1} unless f is injective. We recall that if no two elements of X are mapped into the same element of Y, then f is called *one-to-one or injective*. If every element of Y is the image of at least one element of X, then f is called *onto or surjective*. A mapping which is both injective and surjective is called *bijective*. Then, f^{-1} is defined at all points of Y and there exists a mapping $f^{-1} : Y \to X$ called the *inverse* of f. When f is bijective f^{-1} is also bijective. All things considered, a bijective mapping between two finite sets is a one-to-one correspondence, with both sets having the same number of elements. The bijective mappings f and f^{-1} satisfy $f o f^{-1} = Id_Y$ and $f^{-1} o f = Id_X$ where Id is the identity mapping which takes every element to itself [Mar 91, Sch 80].

A mapping $f : X \to Y$, between metric spaces, is *continuous* at $x \in X$ if for every neighborhood $N \subset Y$ of $f(x)$ there is a neighborhood M of x such that $f(M) \subset N$. A mapping is continuous on X if it is continuous at each $x \in X$.

Theorem. The mapping $f : X \to Y$ is continuous iff for every open set V of Y, the *inverse image* of V, $U = f^{-1}(V) = \{x \in X | f(x) \in V\}$, is an open set of X.

Proof: First suppose that f is continuous and let V be open in Y. Let $x \in f^{-1}(V)$. As V is a neighborhood of $f(x)$, we must have a neighborhood M of x such that $f(M) \subset V$. Consequently, $M \subset f^{-1}(V)$ and any $x \in f^{-1}(V)$ has a neighborhood contained in $f^{-1}(V)$. So, $U = f^{-1}(V)$ is a neighborhood of everyone of its points and as a result $U = f^{-1}(V)$ is open in X. That is, if f is continuous the inverse image of an open set of Y is an open set of X. Conversely, suppose that f satisfies the property that for any open set V of Y, $U = f^{-1}(V)$ is an open set of X. We must now show that f is continuous at any arbitrary point x in X. Let N be a neighborhood of $f(x)$ which contains an open set V containing $f(x)$. So, $f^{-1}(N)$ is a neighborhood containing the assumed open set $U = f^{-1}(V)$ which is a neighborhood of x such that the image of any point in it lies in N. This shows that f is continuous. ∎

Note. There are situations in which, under a continuous mapping, the image of an open set in X is not open in Y. Let $f : \mathbb{R} \to \mathbb{R}$ be a mapping defined by $x \mapsto x^2$. Then f is continuous but the image under f of an open set of $\mathbb{R}$ is not necessarily open in $\mathbb{R}$. For example, an interval such as $(-1, +1)$, open in $\mathbb{R}$, is mapped to $[0, 1)$ which is not an open interval of $\mathbb{R}$. This is so because 0 is not an interior point, i.e., any open ball about 0 contains points of $\mathbb{R}$ which are not in $[0, 1)$. And neither is it closed, because it does not contain all its limit points. Given a subset A of X, $x \in X$ is a limit point of A if every neighborhood of x contains at least some other point in A. ■

In order to preserve the open set structure under "stretching and bending" of the space, and to extend the concept of open set to more general spaces than metric spaces, we need to establish a more general definition, without recourse to the notion of nearness. In this way, open sets appear as members of conveniently defined families called the topologies. A *topology* on a space X is a collection $\mathcal{T}$ of subsets of X satisfying:

- X and the empty set $\emptyset$ are in $\mathcal{T}$
- The union of any number of open sets is in $\mathcal{T}$
- The intersection of a finite number of open sets is in $\mathcal{T}$.

If $\mathcal{T}$ is a topology for X, the members of $\mathcal{T}$ are called *open sets*. Generalizing the notion of metric space, a *topological space* $(X, \mathcal{T})$ consists of a set X for which a topology $\mathcal{T}$ has been specified.

Suppose that $\mathcal{T}$ and $\mathcal{T}'$ are two topologies for the same set X. If $\mathcal{T} \subset \mathcal{T}'$, with unions of open sets in $\mathcal{T}'$ being open sets in $\mathcal{T}$, one says that $\mathcal{T}'$ is finer than $\mathcal{T}$.

Example. Let X be a set containing just two points a and b. There are four topologies on X, which are $\{X, \emptyset\}$, $\{X, \emptyset, \{a\}\}$, $\{X, \emptyset, \{b\}\}$ and $\{X, \emptyset, \{a\}, \{b\}\}$. ■

Every metric space can be thought of as a topological space, regarding the topology as the set of all open sets of the metric space defined in the sense of open balls (1.2). Whenever $\mathbb{R}^n$, the set of all ordered n-tuples of real numbers, is referred to as a metric space, the usual topology of open balls, defined by the Euclidean metric, will be assumed. Clearly, a single point set in $\mathbb{R}^n$ such as $\{p\}$ is not an open set in this topology; regardless of how small an open ball centered at p is, it contains points of $\mathbb{R}^n$ not contained in $\{p\}$.

The notion of continuity was introduced to express the possibility of deforming one space into another continuously, i.e., without tearing. In the framework of metric spaces we started by formulating continuity at a point in terms of nearness. In a more general context, continuity is formulated with respect to specified topologies, with theorems in a metric space becoming definitions in a topological space. The mapping

$$f : (X, \mathcal{T}) \rightarrow (Y, \mathcal{T}') \tag{1.5}$$

is said to be *continuous* if the inverse image of each open set of $(Y, \mathcal{T}')$ is an open set of $(X, \mathcal{T})$.

A *homeomorphism* of topological spaces is a bijective mapping such that f and f^{-1} are both continuous. From the continuity of f and f^{-1}, a homeomorphism sets up a one-to-one correspondence between open sets of $(X, \mathcal{T})$ and open sets of $(Y, \mathcal{T}')$.

For many applications of topological spaces it is convenient to impose a separability condition. A topological space satisfies the *Hausdorff condition* if given any two distinct points $x_1, x_2 \in X$, one may find in $\mathcal{T}$ open sets U_1 and U_2 such that $U_1 \cap U_2 = \emptyset$, with $x_1 \in U_1$ and $x_2 \in U_2$.

Example. Any metric space is a Hausdorff space; the reason being that if $x, y \in X$ and $\epsilon = \frac{1}{2}d(x, y)$, $B(x, \epsilon)$ and $B(y, \epsilon)$ are disjoint open sets. ∎

Chapter 2

Metric Tensor

A *vector space* V is a set of objects (vectors) $\{v_i\}$ with the following properties:

1. Addition (which is commutative and associative).

The set V contains the zero vector 0 and for every vector $v \in V$ there exists a vector $-v$ such that $v + (-v) = 0$. These are conditions satisfied by an additive group.

2. Multiplication by scalars (which is associative and distributive with respect to vector addition and scalar addition).

The multiplication by the scalar 1 leaves the vectors unchanged. The vector space V is called a real vector space or a complex vector space in accordance with the scalars being real or complex numbers.

An *affine space* is a set $\mathcal{V}$ of points, all with equal status, neither having properties of addition and multiplication by scalars nor metric, but on which an n-dimensional real vector space V acts *simply transitively* as a map $V \times \mathcal{V} \to \mathcal{V}$. This means that for each pair of points p and q in $\mathcal{V}$, there is associated an unique vector v in V which maps p on q. The operation on individual elements is described by $(v, p) \mapsto p + v$, with the point to which p is mapped being denoted by $q = p + v$. Since an additive expression such as $q = p + v$ makes sense in affine spaces, it is also meaningful to speak for every pair of ordered points (p, q) of an associated vector in V. As we shall point out in Chapter 3.2, there is nothing analogous in arbitrary manifolds M [CP86, GS87], where p and q are the locations for completely independent vector spaces $T_p M$ and $T_q M$. These vector spaces are separately isomorphic to $\mathbb{R}^n$ and therefore to each other. But the isomorphism, depending on the choice of bases, is not canonical and they cannot be viewed as the "same" in the light of vector spaces. On the other hand, we can consider identical realizations of the associated vector space at different points of an affine space.

A set of linearly independent vectors $e_1, \ldots, e_n$ is called a *basis* of the vector space V if every vector $v \in V$ can be written as a linear combination

$$v = x^1 e_1 + \cdots + x^n e_n \,. \tag{2.1}$$

As by assumption the e_i are linearly independent, the set of scalars x^i is unique. The vector space V is n-dimensional if it contains a set of n linearly independent vectors, but every set of $n + 1$ vectors is linearly dependent.

Choosing an arbitrary point p in $\mathcal{V}$ to play the role of origin $\mathcal{O}$ and a basis $\{e_i\}$ for the vector space V, a point q, in $\mathcal{V}$, may be written as

$$q = \mathcal{O} + v = x^1 e_1 + \cdots + x^n e_n \,. \tag{2.2}$$

The *affine coordinates* of q, with respect to the basis $\{e_i\}$, are defined as the components $x^1, \ldots, x^n$ of the vector $v \in V$ associated with the pair of points $\mathcal{O}$ and q. The assignment of affine coordinates $x^1, \ldots, x^n$ to each point in $\mathcal{V}$ can be described as a bijective mapping between $\mathcal{V}$ and the space of all n-tuples of real numbers $\mathbb{R}^n$. What is lacking is the possibility of comparing the lengths of the basis vectors $e_1, \ldots, e_n$ along different axes. This implies that we cannot ascertain a common measure to the the components $x^1, \ldots, x^n$. An example, taken from thermodynamics, is the (P, V, T) diagram for a gas, where we can represent points, curves, and surfaces, but where the distance between two points is meaningless. In order to introduce a metric property such as length, we need to ascribe a meaning to the norm $\|v\|$ of vectors in different directions. This requires a metric structure to be given to the vector space V.

Previously, metric was defined for an arbitrary set (Chapter 1), but here is defined specifically on a vector space. Let V be an

n-dimensional *real vector space*. A bilinear map

$$g : V \times V \to \mathbb{R} \qquad (2.3)$$

which is a covariant tensor of degree 2, symmetric and non-degenerate, is called a *metric tensor*. The metric tensor makes any pair of vectors $u, v \in V$ correspond to a real number called the *inner product* (*scalar product*)

$$g : (u, v) \mapsto g(u, v) = (u|v) \,. \qquad (2.4)$$

The *norm* of a vector u is defined as $\|u\| = g(u, u) = (u|u)$.

The symmetry requirement means that $g(u, v) = g(v, u)$ for all $u, v \in V$. The non-degeneracy requirement means that $g(u, v) = 0$, for all $v \in V$, implies $u = 0$, that is, if $g(u_1, v) = g(u_2, v)$ for all $v \in V$, then $u_1 = u_2$.

Once a basis $e_1, \ldots, e_n$ has been chosen, the components

$$g_{ij} = g(e_i, e_j) \qquad (2.5)$$

form an $n \times n$ symmetric matrix (g_{ij}) and the inner product can be represented by

$$g(u, v) = g_{ij} u^i v^j \,. \qquad (2.6)$$

Note. By a convenient choice of the Cartesian basis in 3-dimensional Euclidean space, we easily find that the inner product ("dot product") can be represented by the product of the lengths of the two vectors by $\cos \theta$, with θ being the angle between u and v. ∎

With these remarks in hand it can be shown that the matrix (g_{ij}) besides being symmetric has non-vanishing determinant

$$\det(g_{ij}) \neq 0 \,. \qquad (2.7)$$

In fact, the non-degeneracy requirement means that

$$g(u, e_j) = \sum_i u^i g(e_i, e_j) = 0 \qquad (2.8)$$

for all $e_j (j = 1, \ldots, n)$, implies $u^1 = \cdots = u^n = 0$. Letting j take values from 1 to n, this condition is equivalent to a homogeneous system of n linear equations in the n unknowns u^i with *only* the trivial solution $u^1 = \cdots = u^n = 0$. Therefore $\det(g_{ij}) \neq 0$. Thus, (g_{ij}) has an inverse denoted by (g^{ij}) (Chapter 4).

If the vector space associated with $\mathcal{V}$ is given a metric structure, the affine space is called an *inner product affine space*.

A metric is called *positive definitive* if $\|u\| = g(u, u) \geq 0$ for all u, and $\|u\| = g(u, u) = 0$ if and only if $u = 0$. An inner product affine space whose associated vector space has defined on it a positive definitive metric is called an *Euclidean space*.

Being a hermitian matrix, (g_{ij}) can be diagonalized and the diagonal elements can be reduced to the form ± 1 by convenient introduction of scale factors on the basis vectors. So, given a metric tensor g on a vector space, there is a basis $e_1, \ldots, e_n$ such that $g(e_i, e_j) = 0$ for $i \neq j$ and $g(e_i, e_i) = \pm 1$ for $\forall i$. Such a basis is said to be *orthonormal* with respect to g.

Although Euclidean space does not favor a specific coordinate system, a choice of an orthonormal coordinate basis leads to computational simplicity. A model of *n-dimensional Euclidean space* is an inner product affine space whose associated vector space is $\mathbb{R}^n$ plus the *Euclidean metric* $\delta = \text{diag}(1, 1, \ldots, 1)$. An orthonormal *coordinate* basis (Chapter 3.2) in Euclidean space is called *Cartesian*.

The usual procedure of persistently considering $\mathbb{R}^n$ as the vector space of all ordered n-tuples of real numbers $(x^1, \ldots, x^n)$, conveniently associated with a basis $(1, 0, \ldots, 0), \ldots, (0, \ldots, 0, 1)$ and endowed with inner product, gives rise to the habitual practice of denoting the Euclidean space by $\mathbb{R}^n$.

As a topological space, $\mathbb{R}^n$ is referred to as the set of all ordered n-tuples $(x^1, \ldots, x^n)$ of real numbers with the Euclidean distance between $x = (x^1, \ldots, x^n)$ and $y = (y^1, \ldots, y^n)$ defined by

$$d(x, y) = \sqrt{\sum_{i=1}^{n} (x^i - y^i)^2}. \tag{2.9}$$

The topology is defined in the sense of open balls. This is no more than the result obtained from the norm $g(x - y, x - y) = \|x - y\|$ of the vector $x - y$, as we find from

$$g(x - y, x - y) = \sum_{i,j=1}^{n} \delta_{ij}(x^i - y^i)(x^j - y^j). \qquad (2.10)$$

On the other hand, *Minkowski space* is a 4-dimensional inner product affine space whose associated vector space $\mathbb{R}^4$ has coordinates x^0, x^1, x^2, x^3 and *Minkowski metric* $\eta = \text{diag}\,(1, -1, -1, -1)$. A topology for Minkowski space cannot come from the Minkowski metric, which is not definite positive and therefore does not define a distance function [AP95] (Chapter 1). The Minkowski metric allows null separations between distinct points on the path of a light signal and even imaginary values for spacelike separations.

Chapter 3

Differentiable Manifolds

3.1 Basic definitions

An *n-dimensional manifold* M is a Hausdorff topological space locally homeomorphic to $\mathbb{R}^n$. This means that M can be described locally by coordinates attributed by means of a chart around each point [AM78, AMR 88, AP95, BT87, Boo86, CBDMDB91, CP86, CM85, Fra04, GS87, Ish89, Mar91, Nak90, NS83, Sch80, von W81].

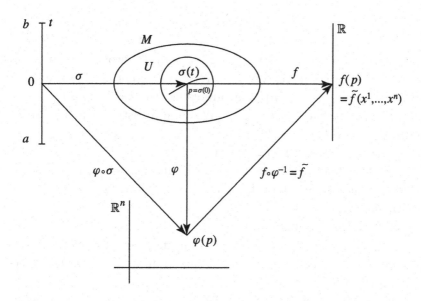

Fig. 3.1   Presentation of curves and functions.

A *chart* (U, φ) on M comprises an open set U of M and a homeomorphism φ of U into an open subset of $\mathbb{R}^n$.

Coordinates are assigned to the image $\varphi(p)$ on $\mathbb{R}^n$ by means of the *coordinate functions* $\pi^i : \mathbb{R}^n \to \mathbb{R}$, such that (Fig. 3.2) $\pi^i o\varphi(p) = x^i(p)$. The so-called coordinate systems (Cartesian, polar etc.) are actually systems of coordinate functions.

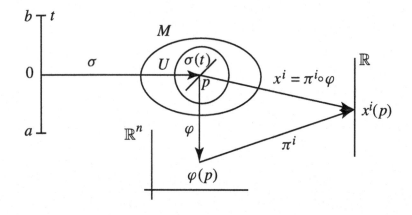

Fig. 3.2   Coordinate functions and local coordinates.

The real-valued functions $x^i = \pi^i o\varphi$ map U into $\mathbb{R}$ and their values at $p \in U \subset M$, defined by $p \mapsto x^i(p)$, represent a set of n ordered numbers $(x^1(p), \ldots, x^n(p))$ that are called the *local coordinates* of p with respect to the chart (U, φ). We should distinguish the two alternatives of $x^1(p), \ldots, x^n(p)$ as coordinates assigned to $\varphi(p)$ in $\mathbb{R}^n$ and as coordinates of p on $U \subset M$ [GS87].

In $\mathbb{R}^n$ we may use Cartesian coordinates or any other basis. In addition, the assumption of a metric on $\mathbb{R}^n$ does not demand any metric on M.

Locally, the manifold brings to mind an affine space but there is nothing relating a point p with another point q by means of a displacement vector represented on an underlying vector space (Chapter 3.2).

Let $f : U \to \mathbb{R}$ be a real-valued function defined on an open set U of M as in Fig. 3.1. With $\tilde{f}$ defined by $f o\varphi^{-1}$, a coordinate value $\tilde{f}(x^1, \ldots, x^n)$, at $\varphi(p)$, can be made to correspond to $f(p)$. The function f is said to be smooth, i.e., infinitely differentiable (C^∞) at $p \in U$, if $\tilde{f}$ is C^∞ at $\varphi(p) \in \mathbb{R}^n$. Basically, we use the differential structure of $\mathbb{R}^n$ to define the differential structure of the manifold. We recall that a function $f : U \to \mathbb{R}$ is said to be C^k-differentiable on U if the partial derivatives of all orders, less than or equal to k, of $\tilde{f} : \varphi(U) \to \mathbb{R}$ exist and are continuous. C^0 means simply continuous.

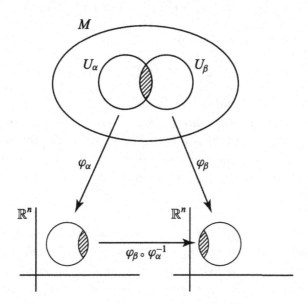

Fig. 3.3 Change of coordinates from $\varphi_\alpha(U_\alpha \cap U_\beta)$ to $\varphi_\beta(U_\alpha \cap U_\beta)$.

In general, it is not possible to find a single chart covering the whole of the manifold. That is, the manifold is not globally homeomorphic to $\mathbb{R}^n$. In this case it is necessary to introduce a collection of charts which give rise to a collection of local coordinate systems. Thus, in the overlap of two charts we have to establish a criterion of compatibility of coordinates. The idea is to introduce additional structure so that the change of coordinates is given by differentiable functions. Suppose that $(U_\alpha, \varphi_\alpha)$ and (U_β, φ_β) are two overlapping charts on M. Since the maps φ_α and φ_β are homeomorphisms, and therefore invertible, we can define a composite homeomorphism

$$\varphi_\beta o\varphi_\alpha^{-1} : \quad \varphi_\alpha(U_\alpha \cap U_\beta) \to \varphi_\beta(U_\alpha \cap U_\beta) \qquad (3.1)$$

which describes how distinct charts are to be pasted together. To each $p \in U_\alpha \cap U_\beta$ we associate, in the charts $(U_\alpha, \varphi_\alpha)$ and (U_β, φ_β) around p, two sets of coordinates $x^i(p)$ and $y^i(p)$, and $\varphi_\beta o\varphi_\alpha^{-1}$ corresponds to a change of coordinates

$$y^j = y^j(x^1, \ldots, x^n) \qquad (j = 1, \ldots, n). \qquad (3.2)$$

Identically, for the inverse map

$$\varphi_\alpha o\varphi_\beta^{-1} : \quad \varphi_\beta(U_\alpha \cap U_\beta) \to \varphi_\alpha(U_\alpha \cap U_\beta). \qquad (3.3)$$

The charts $(U_\alpha, \varphi_\alpha)$ and (U_β, φ_β) are said to be C^∞-related if the homeomorphisms $\varphi_\beta o\varphi_\alpha^{-1}$ and $\varphi_\alpha o\varphi_\beta^{-1}$ are C^∞ (i.e., smooth). If the charts comply with this differentiable structure the coordinate transformation functions have partial derivatives for all orders.

An *atlas* on a manifold M is a family of charts $\{(U_\alpha, \varphi_\alpha)\}$ of M selected in such a way that $\cup U_\alpha = M$ and every pair of charts is C^∞-related.

A *differentiable manifold* is a manifold with an atlas defined over it. Essentially, differentiable manifolds are mathematical objects that globally might not look like $\mathbb{R}^n$, but it makes sense to consider them as being covered by patches that "resemble" $\mathbb{R}^n$ and generate several local coordinate systems. Furthermore, if two coordinate systems are patched together, it is essential that they are related to each other, in the overlapping domain, in a smooth manner. This differentiable structure ensures that the meaning of a differentiable function or map in one coordinate system is consistent when referred to the other. A trivial example of differentiable manifold is $\mathbb{R}^n$, covered by a globally defined single chart $(\mathbb{R}^n, \varphi)$ corresponding to a single coordinate system. Here φ is the identity map $I : \mathbb{R}^n \to \mathbb{R}^n$. Any open subset U of $\mathbb{R}^n$ inherits this property.

A bijective C^∞ mapping between two differentiable manifolds, $F : M \to N$, with a C^∞ inverse F^{-1}, is a *diffeomorphism*. A diffeomorphism is obviously a homeomorphism (C^o).

Diffeomorphisms perform the same role in relation to differentiable manifolds as homeomorphisms perform in relation to topological spaces and isomorphisms in relation to vector spaces or groups in general.

3.2 Tangent vectors and spaces

On an arbitrary manifold M, the idea of vector associated with a pair of points is replaced by the concept of vector referred to some point $p \in M$ where it is located.

As we shall comment later on, vector spaces at distinct points are in general independent of each other, even if the points belong to the same chart.

All the n-dimensional vector spaces are *isomorphic* to each other, i.e., there exists a *bijective correspondence* between elements of one vector space and elements of another, in such a way that the vector space structure under addition and multiplication by scalars of corresponding elements, is preserved. For example, although $\mathbb{C}$ and $\mathbb{R}^2$ differ in the nature of their elements, in terms of their properties as vector spaces $\mathbb{C}$ and $\mathbb{R}^2$ are isomorphic. This results from the identification of the set of all complex numbers $\mathbb{C}$ with points in the complex plane. An isomorphism whose realization does not depend on a choice of basis for its description is said to be a *canonical (uniquely defined) isomorphism*. For example, any two n-dimensional vector spaces are isomorphic. But, this isomorphism, depending in general on the particular choice of bases, is not canonical [AP95, Boo86]. We shall comment in Chap. 4, the consequences of the presence of a metric when considering the mapping between $T_p M$ and $T_p^* M$. Likewise, there is no canonical isomorphism of tangent spaces on all of the two-sphere S^2. Due to well known coordinate singularities [BT87, Nak90, Sch80] no single chart may be used everywhere on S^2. At least two charts are needed and more than one atlas of charts can be used to cover all of S^2. No uniquely defined isomorphism can be fixed between tangent spaces everywhere on S^2 and $\mathbb{R}^2$.

An arbitrary manifold M is not an affine space with associated vector space endowed with metric (Chapter 2). So, we cannot, as in Euclidean space, rely on the limit of the ratio of chord to arc length to define a tangent vector,

$$\hat{t} = \lim_{\delta s \to 0} \frac{\delta p}{\delta s} \tag{3.4}$$

to the curve at p. Nevertheless, this notion associating a vector with a rate of change along a curve is open to generalization in the case of manifolds.

We recall that a *curve* σ on a manifold M is a C^∞ map from an interval $(a,b) \in \mathbb{R}$ (containing $t = 0$) into M, such that the image under σ of $t \in (a,b)$ corresponds to $\sigma(t)$ in M. The map $\sigma : (a,b) \to M$, by $t \in (a,b) \mapsto \sigma(t) \in M$, is said to be C^∞ if for every chart (U, φ) on M, $\varphi o \sigma$ is a C^∞ map.

On a manifold M every vector can be defined to be a tangent vector to a curve in M. Restricting a smooth function f to the neighboring points of $p = \sigma(0)$ that lie on $\sigma(t)$, the composite function $f o \sigma$, from $(a,b) \subset \mathbb{R}$ to $\mathbb{R}$, is a smooth function on $\mathbb{R}$ whose value at t is given by $(f o \sigma)(t) = f(\sigma(t))$. The derivative $\frac{d}{dt}(f o \sigma)(t)\big|_{t=0}$ measures the t rate of change of f along $\sigma(t)$ at $t = 0$ and depends on the curve via the direction of $\sigma(t)$ at $p = \sigma(0)$. Considering two curves σ_1 and σ_2 such that $\sigma_1(0) = \sigma_2(0) = p$, we say that the two curves have the same tangent at p if for any function f

$$\frac{d}{dt}f(\sigma_1(t))\bigg|_{t=0} = \frac{d}{dt}f(\sigma_2(t))\bigg|_{t=0}. \tag{3.5}$$

Therefore, one way to view the *tangent vector* v_p to a curve on a manifold M at p, is as a directional derivative operating on f along $\sigma(t)$ at $t = 0$

$$v_p f := \frac{d}{dt}f(\sigma(t))\bigg|_{t=0} \in \mathbb{R} \tag{3.6}$$

independently of the choice of coordinates. The derivative is taken not in relation to distance between nearby points (which is not defined on the manifold) but with respect to the parameter t, being (3.6) a generalization to a manifold of a directional derivative in $\mathbb{R}^n$.

What is relevant is not the particular function f chosen, but what the directional derivative operator tells us about how the curve causes the function to change from $p = \sigma(0)$.

The set of all tangent vectors at p forms a real vector space denoted by T_pM.

The property (3.5) establishes an equivalence relation among tangent curves at p, and it is possible to identify v_p with an equivalence class $[\sigma]$ of curves tangent at p.

Once selected a representative σ of the equivalence class $[\sigma]$, if $\sigma(t)$ passes through a chart (U, φ) the points on it are represented parametrically (Fig. 3.2) by n real-valued functions $x^i(\sigma(t)) = (x^i o \sigma)(t)$.

So, using local coordinates, we can write for all differentiable functions f at $p = \sigma(0)$

$$v_p f = \left. \frac{d}{dt} f(\sigma(t)) \right|_{t=0} = \left. \frac{\partial f}{\partial x^i} \right|_p \left. \frac{dx^i(\sigma(t))}{dt} \right|_{t=0}. \tag{3.7}$$

With $\tilde{f} = f o \varphi^{-1}$ (Chapter 3.1) we define $\left. \frac{\partial f}{\partial x^i} \right|_p$ by

$$\left. \frac{\partial f}{\partial x^i} \right|_p := \left. \frac{\partial \tilde{f}}{\partial x^i} \right|_{\varphi(p)} \tag{3.8}$$

and the *components* v^i of the tangent vector v_p to $\sigma(t)$, at $p = \sigma(0)$, can be written, in the usual sense, in terms of the local coordinates

$$v^i = \left. \frac{d}{dt} x^i(\sigma(t)) \right|_{t=0}. \tag{3.9}$$

To all curves in the equivalence class $[\sigma]$ of curves tangent at p correspond the same set of values for v^i.

Then, for all smooth functions at p

$$v_p f = v^i \left. \frac{\partial f}{\partial x^i} \right|_p \in \mathbb{R}. \tag{3.10}$$

If v_p is applied to the coordinate functions x^j

$$v_p x^j = v^i \frac{\partial x^j}{\partial x^i} = v^j.$$

In other words, since (3.10) holds for all smooth functions at $p = \sigma(0)$, any *tangent vector* to M along $\sigma(t)$ at $p = \sigma(0)$ can be viewed as a derivation, on the set of all smooth functions, into $\mathbb{R}$

$$v_p = v^i \frac{\partial}{\partial x^i}\bigg|_p . \qquad (3.11)$$

In a coordinate system (U, φ), the map φ assigns coordinates $x^1, \ldots, x^n$ to a point p in U and every mapping $\frac{\partial}{\partial x^i}\big|_p$ is defined by the action on smooth functions by way of $\frac{\partial}{\partial x^i}\big|_p : f \mapsto \frac{\partial \tilde{f}}{\partial x^i}\big|_{\varphi(p)}$. This definition characterizes a directional derivative along the coordinate line on which x^i changes with all the other coordinates remaining fixed. In terms of local coordinates (depending on the choice of a particular chart) the set of vectors

$$\frac{\partial}{\partial x^1}\bigg|_p, \ldots, \frac{\partial}{\partial x^n}\bigg|_p \qquad (3.12)$$

spans the vector space $T_p M$ and is linearly independent. In fact, if there exist $v^i \in \mathbb{R}$ such that $v_p = v^i \frac{\partial}{\partial x^i}\big|_p = 0$, then $v_p x^j = v^j = 0$ for all j. The set $\{\frac{\partial}{\partial x^i}\big|_p\}$ is called a *coordinate basis* or *natural basis* and defines a frame on U. In general, this frame cannot be extended to the whole of M (Chapter 3.3).

Let the point p vary over some open set U. A vector field v on U is a smooth assignment of a tangent vector v_p of $T_p U$ to each point p of U. A vector field is called smooth if operating on every smooth function f produces another smooth real-valued funcion vf whose value at p is given by

$$(vf)(p) := v_p f . \qquad (3.13)$$

From (3.11) we have, in the frame of local coordinates,

$$v_p = v^i(p) \frac{\partial}{\partial x^i}\bigg|_p \quad \text{for} \quad p \in U$$

with the vector v_p being defined by its n components. Hence, a collection of these vectors, with smoothly varying components in the domain U of the chart, is called a vector field on U [GS87, Ish89]

$$v = v^i \frac{\partial}{\partial x^i}. \tag{3.14}$$

A vector field on M is a set of vector fields on every chart of an atlas $\{(U_\alpha, \varphi_\alpha)\}$ such that, if (U, φ) and (U', φ') are charts with the property $U \cap U' \neq \emptyset$ and coordinates x and x'

$$v = v^i \frac{\partial}{\partial x^i} = v'^j \frac{\partial}{\partial x'^j}. \tag{3.15}$$

As the associated frames are related by

$$\frac{\partial}{\partial x'^j} = \frac{\partial x^i}{\partial x'^j} \frac{\partial}{\partial x^i} \tag{3.16}$$

we have, by linear independence of the set of basis vectors,

$$v^i = v'^j \frac{\partial x^i}{\partial x'^j}. \tag{3.17}$$

We are now in the position to carry a little further the previous remark that tangent spaces T_pM and T_qM at distinct points p and q on M are independent of each other, even if these points lie in the same chart. To clarify this point, let us consider two tangent vectors $v_p \in T_pM$ and $v_q \in T_qM$ such that p and q are both in the domain U of a chart (U, φ). A relationship between their components, in the coordinate system associated with the chart, would not be invariant under a change of chart. This is so because of the transformation properties of the components under a change of chart. If (U', φ') is another chart, with $U \cap U' \neq \emptyset$, we would get from (3.17)

$$v_p^i = v_p'^j \left(\frac{\partial x^i}{\partial x'^j} \right)_p$$

with the transformation coefficients varying, in general, from point to point. In particular, two vectors at distinct points p and q cannot be added one to the other or be considered equivalent. The two tangent spaces are isomorphic to each other but there is no canonical isomorphism between T_pM and T_qM. All n-dimensional tangent spaces T_pM are isomorphic to $\mathbb{R}^n$, and to every other, but these isomorphisms are not canonical. Their realization depends on the choice of the chart, distinct charts corresponding to separate isomorphisms to $\mathbb{R}^n$.

Let $\{x^i\}$ be coordinates on some open set U of M and $\{e_i\} = \{\frac{\partial}{\partial x^i}\}$ a basis associated with those coordinates. Bases directly induced by coordinates are known as *coordinate, natural, or holonomic bases*. Because partial derivatives commute [Cou59], we have, for all e_i and e_j in any coordinate basis,

$$[e_i, e_j]f = \left[\frac{\partial}{\partial x^i}, \frac{\partial}{\partial x^j}\right] f = 0. \qquad (3.18)$$

To decide if to a given basis $\{e_i\}$ there corresponds a coordinate system such that $\{e_i\} = \{\frac{\partial}{\partial x^i}\}$, we have to ensure that $[e_i, e_j] = 0$ for all e_i and e_j [MTW73]. Most often, bases are *anholonomic, or non-coordinate bases*, with which no coordinates can be associated.

Example. In Euclidean 3-space described by Cartesian coordinates x^1, x^2, x^3, the metric tensor is given by (4.40)

$$g = \sum_{i=1}^{3} dx^i \otimes dx^i. \qquad (3.19)$$

The basis (dx^1, dx^2, dx^3) of 1-forms or covariant vectors (Chapter 4) such that

$$dx^i \left(\frac{\partial}{\partial x^j}\right) = \delta^i_j \qquad (3.20)$$

is the *dual* (Eq. (4.5)) to the orthonormal, holonomic, natural basis $\left(\frac{\partial}{\partial x^1}, \frac{\partial}{\partial x^2}, \frac{\partial}{\partial x^3}\right)$ of contravariant vectors.

When conveniently restricted, other admissible coordinates in Euclidean 3-space are the *spherical polar coordinates* r, θ, φ

$$0 < r < \infty$$
$$0 < \theta < \pi$$
$$0 \leq \varphi < 2\pi$$
$$x^1 = r \sin\theta \cos\varphi$$
$$x^2 = r \sin\theta \sin\varphi$$
$$x^3 = r \cos\theta. \qquad (3.21)$$

The spherical polar coordinates do not cover $\mathbb{R}^3$ completely. The ranges assumed reflect the well-known singularities along the z-axis (Chapter 7).

The related holonomic, orthogonal but not normalized basis is $\left(\frac{\partial}{\partial r}, \frac{\partial}{\partial \theta}, \frac{\partial}{\partial \varphi}\right)$ with corresponding dual basis $(dr, d\theta, d\varphi)$. The change in coordinates from (x^1, x^2, x^3) to (q^1, q^2, q^3) induces a change of cobasis (Chapter 4)

$$dx^i = \frac{\partial x^i}{\partial q^j} dq^j \qquad (q^1 = r, q^2 = \theta, q^3 = \varphi). \tag{3.22}$$

In Chapter 4 we shall point out that the differentials in (3.22) are not to be regarded as infinitesimal increments of the coordinates but as covectors constituting bases for the cotangent space $T_p^* M$.

Now we shall express the metric tensor in spherical polar coordinates

$$\sum_i dx^i \otimes dx^i = \sum_{i,j,k} \frac{\partial x^i}{\partial q^j} \frac{\partial x^i}{\partial q^k} dq^j \otimes dq^k = \sum_{j,k} g_{jk} dq^j \otimes dq^k \tag{3.23}$$

with

$$g_{jk} = \sum_i \frac{\partial x^i}{\partial q^j} \frac{\partial x^i}{\partial q^k}. \tag{3.24}$$

This gives

$$g = dr \otimes dr + r^2 d\theta \otimes d\theta + r^2 \sin^2 \theta d\varphi \otimes d\varphi \tag{3.25}$$

with

$$g_{rr} = g\left(\frac{\partial}{\partial r}, \frac{\partial}{\partial r}\right) = 1$$

$$g_{\theta\theta} = g\left(\frac{\partial}{\partial \theta}, \frac{\partial}{\partial \theta}\right) = r^2$$

$$g_{\varphi\varphi} = g\left(\frac{\partial}{\partial \varphi}, \frac{\partial}{\partial \varphi}\right) = r^2 \sin^2 \theta$$

$$g_{jk} = 0 \quad \text{if} \quad j \neq k. \tag{3.26}$$

A non-coordinate basis, anholonomic but orthonormal everywhere, is $\left(e_r = \frac{\partial}{\partial r}, e_\theta = \frac{1}{r}\frac{\partial}{\partial \theta}, e_\varphi = \frac{1}{r\sin\theta}\frac{\partial}{\partial \varphi}\right)$ which do not all commute with each other. The dual basis (4.5), such that $\epsilon^i(e_j) = \delta^i_j$, is $(\epsilon^r = dr, \epsilon^\theta = rd\theta, \epsilon^\varphi = r\sin\theta d\varphi)$ giving rise to the metric tensor

$$g = \epsilon^r \otimes \epsilon^r + \epsilon^\theta \otimes \epsilon^\theta + \epsilon^\varphi \otimes \epsilon^\varphi \, . \ \blacksquare \qquad (3.27)$$

3.3 Parallelization

An n-dimensional manifold M is *parallelizable* if there exists on it a set of n linearly independent vector fields $e_1, \ldots, e_n$ such that, at each point $p \in M$, the tangent vectors $e_1(p), \ldots, e_n(p)$ form a basis of T_pM. In general no coordinates can be associated with this frame. The set of vector fields is called a *parallelization* of M and the assignment

$$p \mapsto (e_1(p), \ldots, e_n(p)) \qquad (3.28)$$

of a basis set to the various tangent spaces T_pM, depending smoothly on p, is called a *moving frame*.

If M admits a parallelization $e_1, \ldots, e_n$ and (α_i^j) is a non-singular (invertible) real valued differentiable matrix, another set of linearly independent vector fields of $M, e_1', \ldots, e_n'$, is provided in the usual fashion by

$$e_i' = \alpha_i^j e_j \, . \qquad (3.29)$$

Considering a parallelization $e_1, \ldots, e_n$ of M, the vectors $v_p \in T_pM$ and $v_q \in T_qM$ are said to be *parallel with respect to that parallelization*, if the components c^j in $v_p = c^j e_j(p)$ are equal to those in $v_q = c^j e_j(q)$.

Given a manifold M, a moving frame defined on the whole of M will often not exist. On a chart (U, φ) with coordinates $x^1, \ldots, x^n$, the natural basis $\{\frac{\partial}{\partial x^i}\}$ determines a parallelization on U. But, in general, this cannot be extended throughout M.

The existence of a moving frame on all of M is an intrinsic property of the manifold. For example $\mathbb{R}^n$ is parallelizable but S^2 is not.

If M has only a single chart covering the whole of the manifold, it is clearly parallelizable. This is the case of $\mathbb{R}^n$ where it is possible to choose $\{\frac{\partial}{\partial x^i}\}$ as a set of globally defined linearly independent vector fields. However, on the two-sphere S^2 every smooth vector field must vanish at some point [Mun00] and two vector fields cannot be linearly independent there.

As parallelism is defined with respect to a given parallelization [Nak90] and the two-sphere S^2 is not parallelizable, there is no global definition of parallelism on S^2. Note that we are speaking of parallelism independently of particular paths from one point to the other. Parallel-transported vectors would depend on the chosen path and therefore we cannot speak of distant parallelism.

Chapter 4

Metric Dual

Let T_pM be the tangent space of an n-dimensional manifold M at p. The elements of T_pM are called *contravariant vectors* as well as just *vectors*. The vector space of the linear mappings

$$\omega: \quad T_pM \to \mathbb{R} \tag{4.1}$$

is called the dual vector space T_p^*M or *cotangent space* at p. The dual space T_p^*M is a vector space of dimension n and its elements, which map vectors to scalars, are called *covariant vectors, covectors*, or, in the framework of forms (Chapter 6), *1-forms*. The action of $\omega \in T_p^*M$ on $v \in T_pM$ is denoted by $\omega(v)$.

We are now ready to introduce a basis for T_p^*M *dual* to the natural basis $\{\frac{\partial}{\partial x^i}\}$ for T_pM. With a differentiable real-valued function f on M and a point $p \in M$ we may associate a map

$$df : T_pM \to \mathbb{R} \tag{4.2}$$

defined, independently of a basis, by

$$df : v \mapsto vf \tag{4.3}$$

for all $v \in T_pM$. The real number is vf (Eq. (3.6)) and the covector (1-form) $df \in T_p^*M$ defined in this way, is called the *differential* of f at p [CP86].

So, it is as if the 1-form df holds information with respect to the rates of change $\frac{df(\sigma(t))}{dt}\big|_{t=0}$ of f along any curve in the equivalence class represented by the tangent vector v at $p = \sigma(0)$.

From Eqs. (3.10) and (4.3) the action of the covector df on v can be expressed in local coordinates as

$$df(v) = v^i \frac{\partial f}{\partial x^i}\Big|_p . \tag{4.4}$$

Hence, if we choose f to be the coordinate function x^i and $v = \frac{\partial}{\partial x^j}$ we obtain

$$dx^i \left(\frac{\partial}{\partial x^j} \right) = \frac{\partial x^i}{\partial x^j} = \delta^i_j . \tag{4.5}$$

Thus, it is logical to establish the differentials of the coordinates $\{dx^i(p)\}$ as defining a natural basis of $T^*_p M$, *dual* of the natural basis $\{\frac{\partial}{\partial x^j}\big|_p\}$ of $T_p M$. In fact, in terms of Eq. (4.5) and by expressing $\omega = \omega_i dx^i \in T^*_p M$ and $v = v^j \frac{\partial}{\partial x^j} \in T_p M$ we will be able to write the action of ω on v as an inner product

$$\omega(v) = \omega_i v^i . \tag{4.6}$$

The aforementioned expression

$$df(v) = v^i \frac{\partial f}{\partial x^i}\Big|_p \tag{4.7}$$

together with the expression of $v \in T_p M$ as

$$v = v^j \frac{\partial}{\partial x^j}\Big|_p \tag{4.8}$$

imply that df can be written in the dual basis as

$$df = \frac{\partial f}{\partial x^i}\Big|_p dx^i . \tag{4.9}$$

Note. This result suggests a parallel with the familiar expression of the "differential" of a function on $\mathbb{R}^n$, encountered in elementary calculus [Boo86, CP86, MTW73]. But the elementary concept of "differential" is dominated by the notion of principal part of the increment $\delta f = f(x^1 + \delta x^1, \ldots, x^n + \delta x^n) - f(x^1, \ldots, x^n)$ as a number associated to that displacement near $(x^1, \ldots, x^n)$. To show the "parallel" between the concept of df as a 1-form and the notion of "differential", let us consider a differentiable function on $\mathbb{R}^n$. Upon insertion of v in Eq. (4.9) we may write

$$df(v) = \left.\frac{\partial f}{\partial x^i}\right|_p dx^i(v). \tag{4.10}$$

If we now interpreted v as a displacement $\delta x^j \frac{\partial}{\partial x^j}$ on $\mathbb{R}^n$, from $p = (x^1, \ldots, x^n)$ to $p' = (x^1 + \delta x^1, \ldots, x^n + \delta x^n)$, we would obtain

$$df(p' - p) = \left.\frac{\partial f}{\partial x^i}\right|_p \delta x^i. \tag{4.11}$$

This represents the "differential" at p, corresponding to that displacement, which may be interpreted as being associated to a first order change in f. However, the recognition that the differential df is a 1-form at p, which evaluated on a tangent vector to some curve, at p, is capable of assigning the rate of change of f along the curve, is a more exact treatment. ∎

As pointed out before, T_pM and T_p^*M being vector spaces of the same dimension, are isomorphic. But this isomorphism is, in general, not unique, depending on the chosen bases. Different bases determine different isomorphisms and which covector corresponds to a particular vector cannot be anticipated. However, when a metric is present, a *canonical isomorphism* can be set up between T_pM and T_p^*M. To develop this point we shall next review the metric structure of the space T_pM and the induced metric on the vector space T_p^*M.

As we shall see in Chapter 5, the tensor product $T_p^*M \otimes T_p^*M$ is the vector space of bilinear mappings from the Cartesian product space $T_pM \times T_pM$ (which is the set of ordered pairs (u, v) with $u, v \in T_pM$) into the real numbers $\mathbb{R}$

$$T : T_pM \times T_pM \to \mathbb{R}. \tag{4.12}$$

Indeed, the tensor product $T_p^*M \otimes T_p^*M$ is a vector space, with addition and multiplication by scalars defined by

$$(T_1 + T_2)(u, v) = T_1(u, v) + T_2(u, v)$$
$$(\alpha T)(u, v) = \alpha(T(u, v)) \tag{4.13}$$

where $u, v \in T_pM$ and $\alpha \in \mathbb{R}$. An element of the vector space $T_p^*M \otimes T_p^*M$ is called a tensor T of covariant order two (Chapter 5).

In Chapter 2 we introduced metric on an affine space, in terms of the bilinear map

$$g : V \times V \to \mathbb{R} \qquad (4.14)$$

constructed on the associated real vector space V. The metric could equally have been associated to the collection of tangent spaces T_pM, considered as identical realizations of the vector space V. This tangent space metric exhibits the constant nature of the metric over the whole of the affine space. This constancy has to be abandoned when generalizing metric to arbitrary manifolds [CP86]. A manifold M is said to possess a metric structure if a non-degenerate, symmetric tensor g_p of covariant order two, is assigned on each tangent space T_pM, varying smoothly with p.

The *metric tensor* [MTW73] can be seen as a device in which there is room for the introduction of elements of the Cartesian product space $T_pM \times T_pM$

$$g(\cdot, \cdot) : T_pM \times T_pM \to \mathbb{R}. \qquad (4.15)$$

This can be visualized as being comprised of a vector space isomorphism

$$g(\cdot, \cdot) : T_pM \to T_p^*M \qquad (4.16)$$

described by the effect on individual elements $u \in T_pM$ as

$$g(\cdot, \cdot) : u \mapsto g(u, \cdot) \qquad (4.17)$$

and a linear map

$$g(u, \cdot) : T_pM \to \mathbb{R}. \qquad (4.18)$$

For a given $u \in T_pM$ we can regard $g(u, \cdot)$ as an element of T_p^*M associated with u, such that we may expound the *inner product* of two vectors $u, v \in T_pM$ by

$$g(u, \cdot) : v \mapsto g(u, v) \qquad (4.19)$$

without reference being made to a basis. The geometric structure determined by the metric is an intrinsic feature of the vector space and does not depend on the choice of a basis. From Eq. (4.17) to Eq. (4.19), as the metric is non-degenerate (Chapter 2) the mapping between T_pM and its dual T_p^*M is a bijection, independent of a particular choice of basis, giving rise to a canonical isomorphism between T_pM and T_p^*M.

Therefore, by means of the metric tensor g we associate with every element $u \in T_pM$ a particular *metric dual* with respect to g [BT87,CBDMDB91, Sch80] denoted by

$$u^* := g(u, \cdot) \qquad (4.20)$$

image of u in T_p^*M under the canonical isomorphism. The superscript $*$ is referred to as the *metric dual operation*.

In short, the assignment of a metric tensor g to M endows each tangent space T_pM with an *inner product (scalar product)*

$$g(u, v) = u^*(v) = (u|v) = (v|u) = v^*(u) \qquad (4.21)$$

and defines a canonical isomorphism between T_pM and T_p^*M.

Note. We shall now recall briefly the way in which inner product appears in quantum theory. In quantum theory the state of a system is represented mathematically by a state vector, denoted by a ket $|\psi >$ belonging to a *complex* inner product vector space $\mathcal{E}$. To this vector space is associated a dual vector space $\mathcal{E}^*$ which is the space of the mappings of $\mathcal{E}$ into *the field of complex numbers*. The vectors of the dual vector space $\mathcal{E}^*$ are denoted by $< \phi|$ and are called bras. The bras $< \phi|$ stand in a one-to-one correspondence with the kets $|\phi >$. To any ordered pair of kets $|\phi >$ and $|\psi >$ we associate a *complex number* $< \phi|\psi >$ which is the *inner product* resulting from the action of the bra $< \phi|$ on the ket $|\psi >$. Contrary to the definition of inner product in *real* vector space, $< \phi|\psi >$ and $< \psi|\phi >$ are complex conjugate of each other. ∎

The tensor product space $T_p^* M \otimes T_p^* M$ contains elements of the form $T \otimes Q$, where $T, Q \in T_p^* M$, such that

$$T \otimes Q : T_p M \times T_p M \to \mathbb{R}. \tag{4.22}$$

An element of $T_p^* M \otimes T_p^* M$ that can be written in this form, $T \otimes Q$, is called *decomposable*. The action of the real valued map $T \otimes Q$ on the elements (u, v) of the Cartesian product space $T_p M \times T_p M$ is defined by the formula

$$T \otimes Q(u, v) = T(u)Q(v) \tag{4.23}$$

with $u, v \in T_p M$. The tensor product is not commutative.

Once we choose a coordinate basis $\{ \frac{\partial}{\partial x^i} \}$ as basis vectors, the dual basis is $\{ dx^i \}$. In terms of the basis $\{ dx^i \}$ for $T_p^* M$ we can construct [CBDMDB91], for the tensor product space $T_p^* M \otimes T_p^* M$, a basis set $dx^k \otimes dx^l$ defined by Eqs. (4.5) and (4.23)

$$dx^k \otimes dx^l(u, v) = dx^k(u)dx^l(v) = u^k v^l. \tag{4.24}$$

The dimension of $T_p^* M \otimes T_p^* M$ is n^2 and the metric tensor g can be written as a linear combination of n^2 terms (Chapter 5)

$$g = g_{kl} dx^k \otimes dx^l. \tag{4.25}$$

In fact

$$g(u, v) = u^i v^j g_{kl} dx^k \otimes dx^l \left(\frac{\partial}{\partial x^i}, \frac{\partial}{\partial x^j} \right)$$

$$= u^i v^j g_{kl} \delta_i^k \delta_j^l = g_{ij} u^i v^j$$

as expected (Chapter 2).

As g is a non-degenerate symmetric bilinear map $g : T_p M \times T_p M \to \mathbb{R}$ we can write

$$g \left(\frac{\partial}{\partial x^i}, \frac{\partial}{\partial x^j} \right) = g_{kl} dx^k \left(\frac{\partial}{\partial x^i} \right) dx^l \left(\frac{\partial}{\partial x^j} \right) = g_{ij} \tag{4.26}$$

with the numbers g_{ij} satisfying $g_{ij} = g_{ji}$ and $\det(g_{ij}) \neq 0$.

Once we know the components g_{ij} of g, the norm $||v||$ of a vector $v \in T_pM$ is given by

$$||v|| = g(v,v) = g_{ij}v^iv^j .$$ (4.27)

Note. Perhaps it is appropriate, at this time, to make a brief comment [MTW73] on the expression for the *line element (interval)* usually found is special relativity books. Considering two neighboring points x^i and $x^i + \delta x^i (i = 0, 1, 2, 3)$, the four *increments* δx^i may be understood as components of the displacement vector on the basis $\frac{\partial}{\partial x^0}, \frac{\partial}{\partial x^1}, \frac{\partial}{\partial x^2}, \frac{\partial}{\partial x^3}$. The norm of the vector associated with the pair of neighboring points $||\delta s|| = \delta s^2$, corresponds to the insertion of $\delta x^i \frac{\partial}{\partial x^i}$ in Eq. (4.27) resulting in

$$\delta s^2 = g\left(\delta x^i \frac{\partial}{\partial x^i}, \delta x^j \frac{\partial}{\partial x^j}\right) = g_{ij}\delta x^i \delta x^j . \quad \blacksquare$$ (4.28)

With $\{\frac{\partial}{\partial x^i}\}$ as a natural basis of T_pM and $\{dx^i\}$ the dual basis for T_p^*M, we can write

$$u = u^i \frac{\partial}{\partial x^i}$$ (4.29)

and denoting the components of u^* by u_i

$$u^* = u_i dx^i .$$ (4.30)

For arbitrary $v = v^j \frac{\partial}{\partial x^j}$ in T_pM

$$g(u,v) = u^*(v) = u_iv^j dx^i\left(\frac{\partial}{\partial x^j}\right) = u_jv^j.$$ (4.31)

On the other hand

$$g(u,v) = u^iv^j g\left(\frac{\partial}{\partial x^i}, \frac{\partial}{\partial x^j}\right) = g_{ij}u^iv^j .$$ (4.32)

And since this must hold for all v^j

$$u_j = g_{ji}u^i \quad \text{with} \quad g_{ij} = g_{ji}$$ (4.33)

where u^i are the *contravariant components* of u with respect to the basis $\{\frac{\partial}{\partial x^i}\}$ and u_j are the *covariant components* of the metric dual u^* with respect to the basis $\{dx^j\}$.

Denoting by (g^{ij}) the inverse of the matrix (g_{ij})

$$g^{ij}g_{jk} = \delta^i_k \qquad (4.34)$$

it follows that

$$u^k = g^{kj}u_j \quad \text{with} \quad g^{kj} = g^{jk}. \qquad (4.35)$$

Note. As a consequence of the so-called quotient law of tensors [Sok64], if u_j are the components of an arbitrary covector and $g^{kj}u_j$ is known to be the component u^k of a contravariant vector, then g^{kj} are the components of a symmetric tensor of contravariant order 2. ∎

With the metric dual $\frac{\partial^*}{\partial x^i}$, associated with $\frac{\partial}{\partial x^i}$, denoted by

$$\frac{\partial^*}{\partial x^i} = g\left(\frac{\partial}{\partial x^i}, \cdot\right) \qquad (4.36)$$

we obtain for arbitrary $v = v^j \frac{\partial}{\partial x^j}$ in $T_p M$

$$\frac{\partial^*}{\partial x^i}(v) = g\left(\frac{\partial}{\partial x^i}, v\right) = v^j g\left(\frac{\partial}{\partial x^i}, \frac{\partial}{\partial x^j}\right) = g_{ij}v^j = v_i. \qquad (4.37)$$

On the other hand

$$dx^j(v) = v^i dx^j\left(\frac{\partial}{\partial x^i}\right) = v^j \qquad (4.38)$$

and we may write

$$\frac{\partial^*}{\partial x^i} = g_{ij}dx^j$$

$$dx^i = g^{ij}\frac{\partial^*}{\partial x^j}. \qquad (4.39)$$

In Euclidean space described by Cartesian coordinates, $\frac{\partial^*}{\partial x^i}$ is identified with dx^i.

Once introduced an orthonormal basis in Euclidean space $g_{ij} = \delta_{ij}$

$$g = \delta_{ij} dx^i \otimes dx^j = \sum_i dx^i \otimes dx^i \qquad (4.40)$$

and the contravariant components u^i of u become equal (Eq. (4.33)) to the covariant components u_i of u^*. For example, as grad f is the metric dual of df (see example at the end of this Chapter), using a Cartesian basis in Euclidean 3-space

$$df = \frac{\partial f}{\partial x^1} dx^1 + \frac{\partial f}{\partial x^2} dx^2 + \frac{\partial f}{\partial x^3} dx^3$$

$$\text{grad } f = \frac{\partial f}{\partial x^1} \frac{\partial}{\partial x^1} + \frac{\partial f}{\partial x^2} \frac{\partial}{\partial x^2} + \frac{\partial f}{\partial x^3} \frac{\partial}{\partial x^3} \ . \qquad (4.41)$$

In the non-coordinate, orthonormal, spherical basis (Chapter 3.2) we write

$$df = \frac{\partial f}{\partial r} \epsilon^r + \frac{1}{r} \frac{\partial f}{\partial \theta} \epsilon^\theta + \frac{1}{r \sin \theta} \frac{\partial f}{\partial \varphi} \epsilon^\varphi$$

$$\text{grad } f = \frac{\partial f}{\partial r} e_r + \frac{1}{r} \frac{\partial f}{\partial \theta} e_\theta + \frac{1}{r \sin \theta} \frac{\partial f}{\partial \varphi} e_\varphi \ . \qquad (4.42)$$

In the same way that the symmetric tensor of covariant order 2, expressed by $g : T_pM \times T_pM \to \mathbb{R}$, accounts for the map $T_pM \to T_p^*M$, we can define a *metric* g^* on T_p^*M, expressed by

$$g^* : \quad T_p^*M \times T_p^*M \to \mathbb{R} \qquad (4.43)$$

which accounts for the inverse map $T_p^*M \to T_pM$. The *metric* g^* is a symmetric tensor of contravariant order 2 (Chapter 5). The canonical isomorphism induces the definition of an inner product of two 1-forms on T_p^*M by [BT87, CBDMDB91]

$$g^*(u^*, v^*) = g(u, v) \ . \qquad (4.44)$$

By means of the metric g we associated with every element $u \in T_pM$ a unique 1-form $u^* \in T_p^*M$, the *metric dual* of u with respect to g. Now, by means of the metric g^* we invert the map, associating with every 1-form $u^* \in T_p^*M$ an unique metric dual with respect to g^*, denoted by

$$u := g^*(u^*, \cdot) \tag{4.45}$$

image of u^* in T_pM.

Here, we regard the map

$$g^*(u^*, \cdot) : T_p^*M \to \mathbb{R} \tag{4.46}$$

as an element of T_pM described by the effect on individual elements $v^* \in T_p^*M$ by

$$g^*(u^*, \cdot) : v^* \mapsto g^*(u^*, v^*) = u(v^*). \tag{4.47}$$

According to Eqs. (4.21) and (4.47) we can write Eq. (4.44) in the form

$$u(v^*) = u^*(v) = v^*(u). \tag{4.48}$$

From (4.39) we obtain in terms of coordinate bases

$$dx^i \left(\frac{\partial}{\partial x^j} \right) = g^{ik} \frac{\partial^*}{\partial x^k} \left(\frac{\partial}{\partial x^j} \right)$$
$$= \frac{\partial}{\partial x^j}(dx^i) = \delta_j^i. \tag{4.49}$$

The expansion of the contravariant metric

$$g^* := g^{*kl} \frac{\partial}{\partial x^k} \otimes \frac{\partial}{\partial x^l} \tag{4.50}$$

follows naturally, with the components g^{*ij} of the metric g^*, with respect to the associated dual basis $\{dx^i\}$, given by

$$g^{*ij} = g^*(dx^i, dx^j). \tag{4.51}$$

Furthermore, the components g^{*ij} of the metric g^* correspond to the components g^{ij} of the inverse of the matrix (g_{ij}). In fact, recalling Eq. (4.44) we may write

$$g^* \left(\frac{\partial^*}{\partial x^i}, \frac{\partial^*}{\partial x^j} \right) = g \left(\frac{\partial}{\partial x^i}, \frac{\partial}{\partial x^j} \right) \tag{4.52}$$

and taking into account Eq. (4.39) we obtain

$$g_{ik}g^{*kl}g_{jl} = g_{ij} \tag{4.53}$$

that is, the g^{*kl} are the elements of the inverse of the matrix (g_{kl}). Thus, we can write Eq. (4.50) as

$$g^* := g^{kl}\frac{\partial}{\partial x^k} \otimes \frac{\partial}{\partial x^l} \tag{4.54}$$

with

$$g^{ij} = g^*(dx^i, dx^j). \tag{4.55}$$

With Eq. (4.52) in mind, in Euclidean space described by Cartesian coordinates, the orthonormal basis $\frac{\partial}{\partial x^1}, \ldots, \frac{\partial}{\partial x^n}$ naturally induces orthonormality for the corresponding basis of the dual space

$$g^{ij} = g^*(dx^i, dx^j) = \delta^{ij}. \tag{4.56}$$

In Minkowski space described by coordinates $(x^o = ct, x^1, x^2, x^3)$, related orthonormal basis $(\frac{\partial}{\partial x^o}, \frac{\partial}{\partial x^1}, \frac{\partial}{\partial x^2}, \frac{\partial}{\partial x^3})$ and metric $\eta = \mathrm{diag}(1, -1, -1, -1)$, δ^{ij} could be replaced by η^{ij}.

Example. In the presence of a metric, the vector *gradient* [BT87] of a function f is the metric dual of df (Eq. (4.9))

$$\mathrm{grad}\, f = df^*$$

$$\mathrm{grad}\, f = g^{ij}\frac{\partial f}{\partial x^i}\frac{\partial}{\partial x^j}. \tag{4.57}$$

For Euclidean 3-space in polar coordinates r, θ, φ, orthogonal but not normalized,

$$df = \frac{\partial f}{\partial r}dr + \frac{\partial f}{\partial \theta}d\theta + \frac{\partial f}{\partial \varphi}d\varphi \tag{4.58}$$

and with

$$g^{rr} = 1$$

$$g^{\theta\theta} = \frac{1}{r^2}$$

$$g^{\varphi\varphi} = \frac{1}{r^2 \sin^2 \theta}$$

$$g^{ij} = 0 \quad \text{if} \quad i \neq j \tag{4.59}$$

then

$$\operatorname{grad} f = \frac{\partial f}{\partial r}\frac{\partial}{\partial r} + \frac{1}{r^2}\frac{\partial f}{\partial \theta}\frac{\partial}{\partial \theta} + \frac{1}{r^2 \sin^2 \theta}\frac{\partial f}{\partial \varphi}\frac{\partial}{\partial \varphi}. \tag{4.60}$$

Note that, in polar coordinates, $\frac{\partial f}{\partial r}$, $\frac{\partial f}{\partial \theta}$, $\frac{\partial f}{\partial \varphi}$ are not components of $\operatorname{grad} f$. However, in terms of the non-coordinate (anholonomic) orthonormal spherical basis

$$df = \frac{\partial f}{\partial r}\epsilon^r + \frac{1}{r}\frac{\partial f}{\partial \theta}\epsilon^\theta + \frac{1}{r \sin \theta}\frac{\partial f}{\partial \varphi}\epsilon^\varphi$$

$$\operatorname{grad} f = \frac{\partial f}{\partial r}e_r + \frac{1}{r}\frac{\partial f}{\partial \theta}e_\theta + \frac{1}{r \sin \theta}\frac{\partial f}{\partial \varphi}e_\varphi$$

and for Euclidean 3-space in Cartesian coordinates (x^1, x^2, x^3)

$$df = \frac{\partial f}{\partial x^1}dx^1 + \frac{\partial f}{\partial x^2}dx^2 + \frac{\partial f}{\partial x^3}dx^3$$

$$\operatorname{grad} f = \frac{\partial f}{\partial x^1}\frac{\partial}{\partial x^1} + \frac{\partial f}{\partial x^2}\frac{\partial}{\partial x^2} + \frac{\partial f}{\partial x^3}\frac{\partial}{\partial x^3}. \quad \blacksquare$$

Chapter 5

Tensors

A *tensor* T of type (q, r) and of order $q + r$, at a point p on a differentiable manifold M, is a multilinear map from the Cartesian product of q cotangent spaces at p and r tangent spaces at p, to a real number [AM78, AMR88, CBDMDB91, Mar91]

$$T : \underbrace{T_p^* M \times \cdots \times T_p^* M}_{q \text{ factors}} \times \underbrace{T_p M \times \cdots \times T_p M}_{r \text{ factors}} \to \mathbb{R}. \qquad (5.1)$$

Each element of the Cartesian product is an ordered multiplet of covectors and vectors $(\omega^1, \ldots, \omega^q, v_1, \ldots, v_r)$ with $\omega^i \in T_p^* M$ and $v_i \in T_p M$. *Multilinear* means linear in each factor.

The *tensor product space at p*

$$T_p^{(q,r)}(M) = \underbrace{T_p M \otimes \cdots \otimes T_p M}_{q \text{ factors}} \otimes \underbrace{T_p^* M \otimes \cdots \otimes T_p^* M}_{r \text{ factors}} \qquad (5.2)$$

is the vector space (Chapter 4) of dimension n^{q+r} of those multilinear mappings. An element of this space is called a tensor of type (q, r) (q times contravariant and r times covariant).

Properties. The tensor product of a tensor T of type (k, l) and a tensor Q of type (r, s) is a tensor $T \otimes Q$ of type $(k+r, l+s)$ defined by

$$T \otimes Q \ (\omega^1, \ldots, \omega^{k+r}, v_1, \ldots, v_{l+s})$$

$$= T(\omega^1, \ldots, \omega^k, v_1, \ldots, v_l) Q(\omega^{k+1}, \ldots, \omega^{k+r}, v_{l+1}, \ldots, v_{l+s}). \qquad (5.3)$$

This is a generalization of Eq. (4.23). It can be easily verified from the definition that, for any tensors T, Q, S, the tensor product is associative

$$(T \otimes Q) \otimes S = T \otimes (Q \otimes S). \qquad (5.4)$$

Therefore, we simply write $T \otimes Q \otimes S$. The tensor product is also distributive over addition. For tensors Q_1 and Q_2 of the same type

$$T \otimes (Q_1 + Q_2) = T \otimes Q_1 + T \otimes Q_2. \qquad (5.5)$$

However, the tensor product is not commutative, $T \otimes Q \neq Q \otimes T$. ∎

A tensor of type $(0,0)$ is defined to be a scalar, so $T_p^{(0,0)}(M) = \mathbb{R}$. In particular, $T_p^{(0,1)}(M) = T_p^* M$, $T_p^{(1,0)}(M) = T_p M$, and $T_p^{(0,r)}(M)$ is the space of covariant tensors of order r.

Because the symmetries of tensors of type $(0, r)$ are of special importance to applications, let us start by examining their symmetry properties. A tensor T of type $(0, r)$ is a map which takes in the set of vectors $(v_1, \ldots, v_r)$, where $v_i \in T_p M$, and produces a real number $T(v_1, \ldots, v_r)$. Under a permutation P_{jk} of two indices, the symmetry operation is defined by

$$P_{jk} T(v_1, \ldots, v_j, \ldots, v_k, \ldots, v_r) = T(v_1, \ldots, v_k, \ldots, v_j, \ldots, v_r). \qquad (5.6)$$

A covariant tensor is said to be *symmetric* if by interchanging any two of the indices it satisfies

$$T(v_1, \ldots, v_k, \ldots, v_j, \ldots, v_r) = T(v_1, \ldots, v_j, \ldots, v_k, \ldots, v_r). \qquad (5.7)$$

A covariant tensor is *antisymmetric* if it changes sign under an interchange of any two of the indices

$$T(v_1, \ldots, v_k, \ldots, v_j, \ldots, v_r) = -T(v_1, \ldots, v_j, \ldots, v_k, \ldots, v_r). \qquad (5.8)$$

Symmetric and antisymmetric contravariant tensors are defined similarly.

Next we shall see how to express these symmetries in terms of a basis. Taking for definiteness a coordinate basis, we shall show first that the set

$$\left\{ \frac{\partial}{\partial x^{i_1}} \otimes \cdots \otimes \frac{\partial}{\partial x^{i_q}} \otimes dx^{j_1} \otimes \cdots \otimes dx^{j_r}; i_1,\ldots,i_q,j_1,\ldots,j_r=1,\ldots,n \right\}$$

$$(5.9)$$

is a basis of $T_p^{(q,r)}(M)$ with respect to a local coordinate system. We say that the set (5.9) defines a basis of $T_p^{(q,r)}(M)$ if its elements span $T_p^{(q,r)}(M)$ and are linearly independent. Choosing *any* collection $(\omega^1,\ldots,\omega^q,v_1,\ldots,v_r)$ of covector and vector arguments, any tensor $T \in T_p^{(q,r)}(M)$, of contravariant order q and covariant order r, is a multilinear mapping

$$T\left(\omega^1,\ldots,\omega^q,v_1,\ldots,v_r\right)$$

$$= \omega_{i_1}^1 \cdots \omega_{i_q}^q v_1^{j_1} \ldots v_r^{j_r} T\left(dx^{i_1},\ldots,dx^{i_q},\frac{\partial}{\partial x^{j_1}},\ldots,\frac{\partial}{\partial x^{j_r}}\right)$$

$$= T_{j_1\cdots j_r}^{i_1\cdots i_q} \omega_{i_1}^1 \cdots \omega_{i_q}^q v_1^{j_1} \cdots v_r^{j_r}.$$

$$(5.10)$$

The sum runs over $i_1,\ldots,i_q,j_1,\ldots,j_r = 1,\ldots,n$ and

$$T_{j_1\cdots j_r}^{i_1\cdots i_q} = T\left(dx^{i_1},\ldots,dx^{i_q},\frac{\partial}{\partial x^{j_1}},\ldots,\frac{\partial}{\partial x^{j_r}}\right).$$

$$(5.11)$$

To the element $\frac{\partial}{\partial x^{i_1}} \otimes \cdots \otimes \frac{\partial}{\partial x^{i_q}} \otimes dx^{j_1} \otimes \cdots \otimes dx^{j_r}$ of $T_p^{(q,r)}(M)$ corresponds a mapping

$$\frac{\partial}{\partial x^{i_1}} \otimes \cdots \otimes \frac{\partial}{\partial x^{i_q}} \otimes dx^{j_1} \otimes \cdots \otimes dx^{j_r}(\omega^1,\ldots,\omega^q,v_1,\ldots,v_r)$$

$$= \omega_{i_1}^1 \cdots \omega_{i_q}^q \cdots v_1^{j_1} \cdots v_r^{j_r}.$$

$$(5.12)$$

Therefore, we can write (5.10), for *any* $(\omega^1,\ldots,\omega^q,v_1,\ldots,v_r)$, in the form

$$T\left(\omega^1,\ldots,\omega^q,v_1,\ldots,v_r\right)$$

$$= T_{j_1\cdots j_r}^{i_1\cdots i_q} \frac{\partial}{\partial x^{i_1}} \otimes \cdots \otimes \frac{\partial}{\partial x^{i_q}} \otimes dx^{j_1} \otimes \cdots \otimes dx^{j_r}(\omega^1,\ldots,\omega^q,v_1,\ldots,v_r).$$

$$(5.13)$$

So, any tensor T of type (q, r) can be written as

$$T = T^{i_1 \cdots i_q}_{j_1 \cdots j_r} \frac{\partial}{\partial x^{i_1}} \otimes \cdots \otimes \frac{\partial}{\partial x^{i_q}} \otimes dx^{j_1} \otimes \cdots \otimes dx^{j_r} \qquad (5.14)$$

with $T^{i_1 \cdots i_q}_{j_1 \cdots j_r}$ given by (5.11). This proves that (5.9) spans $T^{(q,r)}_p(M)$.

By showing that the assignment of the value zero to (5.14) implies all $T^{i_1 \cdots i_q}_{j_1 \cdots j_r} = 0$, we shall show next that the elements of (5.9) are linearly independent. First of all, the attribution of the value zero to (5.14) is taken to mean that the result of evaluating it on *any* set $(\omega^1, \ldots, \omega^q, v_1, \ldots, v_r)$ is zero. Accordingly, a particular collection $(dx^{k_1}, \ldots, dx^{k_q}, \frac{\partial}{\partial x^{l_1}}, \ldots, \frac{\partial}{\partial x^{l_r}})$ should produce

$$T^{i_1 \ldots i_q}_{j_1 \ldots j_r} \frac{\partial}{\partial x^{i_1}} \otimes \cdots \otimes \frac{\partial}{\partial x^{i_q}} \otimes dx^{j_1} \otimes \cdots$$

$$\cdots \otimes dx^{j_r} \left(dx^{k_1}, \ldots, dx^{k_q}, \frac{\partial}{\partial x^{l_1}}, \ldots, \frac{\partial}{\partial x^{l_r}} \right)$$

$$= 0. \qquad (5.15)$$

Using the result $dx^i \left(\frac{\partial}{\partial x^j} \right) = \frac{\partial}{\partial x^j}(dx^i) = \delta^i_j$ (4.49) it follows that

$$\frac{\partial}{\partial x^{i_1}} \otimes \cdots \otimes \frac{\partial}{\partial x^{i_q}} \otimes dx^{j_1} \otimes \cdots \otimes dx^{j_r} \left(dx^{k_1}, \ldots, dx^{k_q}, \frac{\partial}{\partial x^{l_1}}, \ldots, \frac{\partial}{\partial x^{l_r}} \right)$$

$$= \frac{\partial}{\partial x^{i_1}}(dx^{k_1}) \cdots \frac{\partial}{\partial x^{i_q}}(dx^{k_q}) dx^{j_1} \left(\frac{\partial}{\partial x^{l_1}} \right) \cdots dx^{j_r} \left(\frac{\partial}{\partial x^{l_r}} \right)$$

$$= \delta^{k_1}_{i_1} \cdots \delta^{k_q}_{i_q} \delta^{j_1}_{l_1} \cdots \delta^{j_r}_{l_r} \qquad (5.16)$$

which shows that $\frac{\partial}{\partial x^{i_1}} \otimes \cdots \otimes \frac{\partial}{\partial x^{i_q}} \otimes dx^{j_1} \otimes \cdots \otimes dx^{j_r}$ is non-zero. Therefore, $T = T^{i_1 \cdots i_q}_{j_1 \cdots j_r} \frac{\partial}{\partial x^{i_1}} \otimes \cdots \otimes \frac{\partial}{\partial x^{i_q}} \otimes dx^{j_1} \otimes \cdots \otimes dx^{j_r}$ can only be zero if all the $T^{i_1 \cdots i_q}_{j_1 \cdots j_r}$ are zero. This proves that the set (5.9) is linearly independent. So, (5.9) defines a basis of $T^{(q,r)}_p(M)$ which has dimension n^{q+r}. The numbers $T^{i_1 \cdots i_q}_{j_1 \cdots j_r}$ (5.11) can be rightly called components, for the given basis, in the space $T^{(q,r)}_p(M)$.

Let T be an antisymmetric covariant tensor. The components of T can be written as

$$T\left(\frac{\partial}{\partial x^{j_1}}, \ldots, \frac{\partial}{\partial x^{j_r}}\right) = T_{j_1 \ldots j_r} . \tag{5.17}$$

The symmetry of a tensor is an intrinsic property which can be expressed with respect to an arbitrary basis. Taking the coordinate basis, for any permutation of the indices $j_1 \cdots j_r$ we have

$$PT\left(\frac{\partial}{\partial x^{j_1}}, \ldots, \frac{\partial}{\partial x^{j_r}}\right) = T\left(\frac{\partial}{\partial x^{P(j_1)}}, \ldots, \frac{\partial}{\partial x^{P(j_r)}}\right) = T_{P(j_1) \cdots P(j_r)} \tag{5.18}$$

where P can be expressed as the product of interchange permutations $P_{j_i j_k}$ of two indices. Therefore, the antisymmetry can be expressed by

$$T\left(\frac{\partial}{\partial x^{P(j_1)}}, \ldots, \frac{\partial}{\partial x^{P(j_r)}}\right) = \operatorname{sgn}(P) T\left(\frac{\partial}{\partial x^{j_1}}, \ldots, \frac{\partial}{\partial x^{j_r}}\right) \tag{5.19}$$

or

$$T_{P(j_1) \cdots P(j_r)} = \operatorname{sgn}(P) T_{j_1 \cdots j_r}. \tag{5.20}$$

If the permutation P can be accomplished by the product of an even (odd) number of interchange permutations $P_{j_1 j_k}$ of pairs, we say that $\operatorname{sgn}(P) = +1 (\operatorname{sgn}(P) = -1)$.

Chapter 6

r-Forms

In this chapter we shall give particular attention to the process of antisymmetrization of a tensor. For this purpose we define the *antisymmetrization operator* by

$$\mathcal{A} = \frac{1}{r!} \sum_P \text{sgn}\,(P)P\,. \tag{6.1}$$

In the case of a covariant tensor $T \in T_p^{(0,r)}(M)$ the components of an antisymmetrized covariant tensor can be defined by

$$\mathcal{A}T \left(\frac{\partial}{\partial x^{j_1}}, \ldots, \frac{\partial}{\partial x^{j_r}} \right) = (\mathcal{A}T)_{j_1 \cdots j_r}$$

$$= \frac{1}{r!} \sum_P \text{sgn}\,(P)T \left(\frac{\partial}{\partial x^{P(j_1)}}, \ldots, \frac{\partial}{\partial x^{P(j_r)}} \right)$$

$$= \frac{1}{r!} \sum_P \text{sgn}\,(P)T_{P(j_1)\cdots P(j_r)} \tag{6.2}$$

where the sum is over all permutations.

An r-form τ is an antisymmetric covariant tensor of order r. The antisymmetric tensors in $T_p^{(0,r)}(M)$ form a subspace $\Omega_p^r(M)$ which is the space of r-forms at p.

Our next task is to find out what we can learn from the antisymmetrization, which maps $T_p^{(0,r)}(M)$ onto $\Omega_p^r(M)$, to construct a basis for $\Omega_p^r(M)$ out of the basis

$$\{dx^{j_1} \otimes \cdots \otimes dx^{j_r}; j_i = 1, \ldots, n\} \tag{6.3}$$

for $T_p^{(0,r)}(M)$. To start with, an expression such as

$$\frac{1}{r!} \sum_P \text{sgn}\,(P) \tau_{P(j_1)\cdots P(j_r)} dx^{j_1} \otimes \cdots \otimes dx^{j_r} \tag{6.4}$$

is not an appropriate expansion of an r-form. In fact, as the interchange of any pair of indices should change the sign of the above expression, there should be provision for that expression to vanish if $j_1 \cdots j_r$ are not all distinct. To account for this the *Levi-Civita antisymmetric symbol* $\varepsilon_{j_1\cdots\cdots\cdots j_r}^{P(j_1)\cdots P(j_r)}$ can be introduced. This symbol coincides with sgn (P) if the indices are all distinct and is zero if there are repeated indices. Therefore, an improvement on that expression would be to write

$$\tau = \frac{1}{r!}\, \varepsilon_{j_1\cdots j_r}^{i_1\cdots i_r} \tau_{i_1\cdots i_r} dx^{j_1} \otimes \cdots \otimes dx^{j_r} \tag{6.5}$$

where $\varepsilon_{j_1\cdots j_r}^{i_1\cdots i_r}$ is $+1(-1)$ according to $i_1 \cdots i_r$ is an even (odd) permutation of $j_1 \cdots j_r$, being all indices distinct and otherwise zero, and $\tau_{i_1\cdots i_r} = \tau\left(\frac{\partial}{\partial x^{i_1}}, \ldots, \frac{\partial}{\partial x^{i_r}}\right)$.

We shall next see how to extract an antisymmetric covariant tensor of order $k+r$ from the tensor product of a k-form and an r-form. We first remember that the tensor product $\omega \otimes \tau$ of a k-form ω by an r-form τ, although it is a $(k+r)$-covariant tensor

$$\omega \otimes \tau \left(\frac{\partial}{\partial x^{j_1}}, \ldots, \frac{\partial}{\partial x^{j_k}}, \frac{\partial}{\partial x^{j_{k+1}}}, \ldots, \frac{\partial}{\partial x^{j_{k+r}}} \right)$$

$$= \omega \left(\frac{\partial}{\partial x^{j_1}}, \ldots, \frac{\partial}{\partial x^{j_k}} \right) \tau \left(\frac{\partial}{\partial x^{j_{k+1}}}, \ldots, \frac{\partial}{\partial x^{j_{k+r}}} \right)$$

$$(1 \leq j_1, \ldots, j_k, j_{k+1}, \ldots, j_{k+r} \leq n) \tag{6.6}$$

it is not a $(k+r)$-form. This is because the expression is no longer antisymmetric with respect to permutations exchanging indices in the range $j_1 \cdots j_k$ into the range $j_{k+1} \cdots j_{k+r}$. That is, the tensor product $\otimes$ does not make the set of all forms into an algebra. What we have to do is to extract the antisymmetric part of $\omega \otimes \tau$. We are thus led to define an antisymmetrized product called *wedge product* and denoted by $\wedge$

$$\omega \wedge \tau = \frac{(k+r)!}{k!r!} \mathcal{A}(\omega \otimes \tau) \tag{6.7}$$

which defines a $(k+r)$-form, $\omega \wedge \tau \in \Omega_p^{k+r}(M)$. The numerical factors are introduced for later convenience. As previously in Eq. (6.2), the components of $\mathcal{A}(\omega \otimes \tau)$ are determined by

$$\mathcal{A}(\omega \otimes \tau)\left(\frac{\partial}{\partial x^{j_1}}, \ldots, \frac{\partial}{\partial x^{j_{k+r}}}\right) = (\mathcal{A}(\omega \otimes \tau))_{j_1 \cdots j_{k+r}}$$

$$= \frac{1}{(k+r)!} \sum_P \operatorname{sgn}(P)(\omega \otimes \tau)\left(\frac{\partial}{\partial x^{P(j_1)}}, \ldots, \frac{\partial}{\partial x^{P(j_{k+r})}}\right)$$

$$= \frac{1}{(k+r)!} \sum_P \operatorname{sgn}(P)\omega\left(\frac{\partial}{\partial x^{P(j_1)}}, \ldots, \frac{\partial}{\partial x^{P(j_k)}}\right) \tau\left(\frac{\partial}{\partial x^{P(j_{k+1})}}, \right.$$

$$\left. \ldots, \frac{\partial}{\partial x^{P(j_{k+r})}}\right) \tag{6.8}$$

where $\sum_P$ denotes the sum over all permutations of $j_1, \ldots, j_{k+r}$. In particular, if ω and τ are 1-forms

$$\mathcal{A}(\omega \otimes \tau)\left(\frac{\partial}{\partial x^{j_1}}, \frac{\partial}{\partial x^{j_2}}\right) = (\mathcal{A}(\omega \otimes \tau))_{j_1 j_2}$$

$$= \frac{1}{2} \sum_P \operatorname{sgn}(P)(\omega \otimes \tau)\left(\frac{\partial}{\partial x^{P(j_1)}}, \frac{\partial}{\partial x^{P(j_2)}}\right)$$

$$= \frac{1}{2}\left(\omega\left(\frac{\partial}{\partial x^{j_1}}\right)\tau\left(\frac{\partial}{\partial x^{j_2}}\right) - \omega\left(\frac{\partial}{\partial x^{j_2}}\right)\tau\left(\frac{\partial}{\partial x^{j_1}}\right)\right)$$

$$= \frac{1}{2}(\omega \otimes \tau - \tau \otimes \omega)\left(\frac{\partial}{\partial x^{j_1}}, \frac{\partial}{\partial x^{j_2}}\right). \tag{6.9}$$

This being true for the coordinate basis or any other basis, when ω and τ are 1-forms we have

$$\mathcal{A}(\omega \otimes \tau) = \frac{1}{2}(\omega \otimes \tau - \tau \otimes \omega). \tag{6.10}$$

Consequently, we obtain for the antisymmetrized wedge product of 1-forms

$$\omega \wedge \tau = \omega \otimes \tau - \tau \otimes \omega. \tag{6.11}$$

Obviously, the wedge product of 1-forms is anticommutative

$$\omega \wedge \tau = -\tau \wedge \omega \tag{6.12}$$

and the wedge product of a 1-form with itself vanishes, $\omega \wedge \omega = 0$.

Example. Now we show that, given $\omega \in \Omega_p^k(M)$ and $\tau \in \Omega_p^r(M)$

$$\omega \wedge \tau = (-)^{kr} \tau \wedge \omega. \tag{6.13}$$

In fact, we have for the coordinate basis or any other basis

$$(\omega \otimes \tau) \left(\frac{\partial}{\partial x^{j_1}}, \ldots, \frac{\partial}{\partial x^{j_k}}, \frac{\partial}{\partial x^{j_{k+1}}}, \ldots, \frac{\partial}{\partial x^{j_{k+r}}} \right)$$

$$= (\tau \otimes \omega) \left(\frac{\partial}{\partial x^{j_{k+1}}}, \ldots, \frac{\partial}{\partial x^{j_{k+r}}}, \frac{\partial}{\partial x^{j_1}}, \ldots, \frac{\partial}{\partial x^{j_k}} \right)$$

$$= (-)^{kr} (\tau \otimes \omega) \left(\frac{\partial}{\partial x^{j_1}}, \ldots, \frac{\partial}{\partial x^{j_k}}, \frac{\partial}{\partial x^{j_{k+1}}}, \ldots, \frac{\partial}{\partial x^{j_{k+r}}} \right) \tag{6.14}$$

where the factor $(-)^{kr}$ arises because the permutation is obtained by k transpositions r times. So, $\omega \wedge \tau = (-)^{kr} \tau \wedge \omega$ as asserted in Eq. (6.13). In particular, the wedge product of odd forms is anticommutative and the wedge product of an odd form for itself vanishes. This result is a generalization of expression (6.12). ∎

Properties. We next discuss some properties of the wedge product. From the definition, it is manifest that the wedge product is distributive

$$\omega \wedge (\tau_1 + \tau_2) = \omega \wedge \tau_1 + \omega \wedge \tau_2. \tag{6.15}$$

Before showing that the wedge product is associative we first prove [CM85, Mar91] that given two covariant tensors ω and τ of order k and r, respectively,

$$\mathcal{A}(\mathcal{A}(\omega) \otimes \tau) = \mathcal{A}(\omega \otimes \mathcal{A}\tau) = \mathcal{A}(\omega \otimes \tau). \tag{6.16}$$

Let $\sum_P$ denote the sum over all permutations of $(1, 2, \ldots, k+r)$ and let $\sum_\Pi$ be the sum over permutations acting on $(1, 2, \ldots, k)$ as P does, whilst leaving $(k+1, \ldots, k+r)$ unaffected. We can then write

$$(\mathcal{A}\omega) \otimes \tau = \frac{1}{k!} \sum_\Pi \operatorname{sgn}(\Pi)\Pi(\omega \otimes \tau). \qquad (6.17)$$

Then

$$\mathcal{A}\left((\mathcal{A}\omega) \otimes \tau\right) = \frac{1}{(k+r)!} \sum_P \operatorname{sgn}(P)P\left[\frac{1}{k!} \sum_\Pi \operatorname{sgn}(\Pi)\Pi(\omega \otimes \tau)\right]$$

$$= \frac{1}{(k+r)!}\frac{1}{k!} \sum_\Pi \sum_P \operatorname{sgn}(P)\operatorname{sgn}(\Pi)P\Pi(\omega \otimes r). \quad (6.18)$$

For *each* permutation Π, $P\Pi$ runs through all permutations of $(1, 2, \ldots, k+r)$, as P does, and Eq. (6.18) becomes

$$\frac{1}{(k+r)!} \sum_P \operatorname{sgn}(P\Pi)P\Pi(\omega \otimes \tau) = \mathcal{A}(\omega \otimes \tau) \qquad (6.19)$$

where $\operatorname{sgn}(P\Pi) = \operatorname{sgn}(P)\operatorname{sgn}(\Pi)$. By acting with $\frac{1}{k!}\sum_\Pi$ on the above expression we obtain

$$\frac{1}{k!} \sum_\Pi \frac{1}{(k+r)!} \sum_P \operatorname{sgn}(P\Pi)P\Pi(\omega \otimes \tau) = \frac{1}{k!} \sum_\Pi \mathcal{A}(\omega \otimes \tau). \quad (6.20)$$

As the sum over Π reduces to $k!\mathcal{A}(\omega \otimes r)$ we arrive at

$$\mathcal{A}\left((\mathcal{A}\omega) \otimes \tau\right) = \mathcal{A}(\omega \otimes \tau). \qquad (6.21)$$

Similar calculations lead to

$$\mathcal{A}(\omega \otimes \mathcal{A}\tau) = \mathcal{A}(\omega \otimes \tau). \qquad (6.22)$$

Therefore $\mathcal{A}\left((\mathcal{A}\omega) \otimes \tau\right)$ and $\mathcal{A}(\omega \otimes \mathcal{A}\tau)$ are both equal to $\mathcal{A}(\omega \otimes \tau)$ and this proves Eq. (6.16).

Let ω, τ and σ be forms of order k, r and s, respectively. From Eq. (6.16) and the associativity of the tensor product (Chapter 5) it follows that [CM85]

$$(\omega \wedge \tau) \wedge \sigma = \frac{(k+r+s)!}{(k+r)!s!} \mathcal{A}((\omega \wedge \tau) \otimes \sigma)$$

$$= \frac{(k+r+s)!}{(k+r)!s!} \frac{(k+r)!}{k!r!} \mathcal{A}(\mathcal{A}(\omega \otimes \tau) \otimes \sigma)$$

$$= \frac{(k+r+s)!}{k!r!s!} \mathcal{A}(\omega \otimes \tau \otimes \sigma). \tag{6.23}$$

Similarly,

$$\omega \wedge (\tau \wedge \sigma) = \frac{(k+r+s)!}{k!r!s!} \mathcal{A}(\omega \otimes \tau \otimes \sigma). \tag{6.24}$$

These results, Eqs. (6.23) and (6.24), show that the wedge product is associative

$$\omega \wedge (\tau \wedge \sigma) = (\omega \wedge \tau) \wedge \sigma = \omega \wedge \tau \wedge \sigma$$

$$= \frac{(k+r+s)!}{k!r!s!} \mathcal{A}(\omega \otimes \tau \otimes \sigma). \tag{6.25}$$

In particular, out the natural cobasis $\{dx^i\}$ for $T_p^* M = \Omega_p^1(M)$ we obtain

$$dx^{i_1} \wedge dx^{i_2} \wedge \cdots \wedge dx^{i_r} = r! \mathcal{A}(dx^{i_1} \otimes dx^{i_2} \otimes \cdots \otimes dx^{i_r}). \blacksquare \tag{6.26}$$

Baring in mind the endeavour to construct a first approach to a basis for $\Omega_p^r(M)$, and recalling the comments made after Eq. (6.4), it is worthwhile to write (6.26) in the format

$$dx^{i_1} \wedge \cdots \wedge dx^{i_r} = \varepsilon_{j_1 \cdots j_r}^{i_1 \cdots i_r} dx^{j_1} \otimes \cdots \otimes dx^{j_r} \tag{6.27}$$

where $\varepsilon_{j_1 \cdots j_r}^{i_1 \cdots i_r}$ is $+1(-1)$ according to $j_1 \cdots j_r$ is an even (odd) permutation of $i_1 \ldots i_r$, being all distinct, and otherwise zero. Consequently the only non-zero contributions are those for which $r \le n$, otherwise some of the indices are necessarily repeated.

Example. For $i_1 = 1, i_2 = 2, i_3 = 3$, it follows that

$$dx^1 \wedge dx^2 \wedge dx^3$$
$$= dx^1 \otimes dx^2 \otimes dx^3 + dx^3 \otimes dx^1 \otimes dx^2 + dx^2 \otimes dx^3 \otimes dx^1$$
$$-dx^2 \otimes dx^1 \otimes dx^3 - dx^1 \otimes dx^3 \otimes dx^2 - dx^3 \otimes dx^2 \otimes dx^1. \quad \blacksquare$$

From (6.5) an r-form can be written as

$$\tau = \frac{1}{r!} \tau_{i_1 \cdots i_r} dx^{i_1} \wedge \cdots \wedge dx^{i_r} \tag{6.28}$$

where, since τ is antisymmetric, the coefficients $\tau_{i_1 \cdots i_r} = \tau(\frac{\partial}{\partial x^{i_1}}, \ldots, \frac{\partial}{\partial x^{i_r}})$ change sign under the interchange of any pair of indices and the sum runs over *all* distinct values of $i_1, \ldots, i_r$ in the ranges 1 to n. Nevertheless, the set

$$\{dx^{i_1} \wedge \cdots \wedge dx^{i_r} ; i_k = 1, \ldots, n\} \tag{6.29}$$

is not a basis in the space $\Omega_p^r(M)$ because its elements are not independent, as can be seen from the consideration of expressions like

$$dx^{i_1} \wedge dx^{i_2} \wedge \cdots \wedge dx^{i_r} = -dx^{i_2} \wedge dx^{i_1} \wedge \cdots \wedge dx^{i_r}. \tag{6.30}$$

Therefore, $\frac{1}{r!} \tau_{i_1 \cdots i_r}$ are not components of τ.

Next we shall prove that the set

$$\{dx^{I_1} \wedge \cdots \wedge dx^{I_r} ; 1 \leq I_1 < \cdots < I_r \leq n\} \tag{6.31}$$

defines a *basis* of $\Omega_p^r(M)$. To do so we must show that every r-form τ may be expressed as a linear combination of the proposed wedge products (6.31) and the elements in (6.31) are linearly independent.

First we show that (6.31) spans $\Omega_p^r(M)$. To start with, note that to each specific set of values taken by the indices $i_1, i_2, \ldots, i_r$

corresponds in (6.28) $r!$ identical expressions

$$\frac{1}{r!}\tau_{i_1 i_2 \cdots i_r} dx^{i_1} \wedge dx^{i_2} \wedge \cdots \wedge dx^{i_r}$$

$$= \frac{1}{r!}\tau_{i_2 i_1 \cdots i_r} dx^{i_2} \wedge dx^{i_1} \wedge \cdots \wedge dx^{i_r}$$

$$= \frac{1}{r!}\tau_{P(i_1)P(i_2)\cdots P(i_r)} dx^{P(i_1)} \wedge dx^{P(i_2)} \wedge \cdots \wedge dx^{P(i_r)}$$

since under permutation both $\tau_{i_1 i_2 \cdots i_r}$ and $dx^{i_1} \wedge dx^{i_2} \wedge \cdots \wedge dx^{i_r}$ change by $\text{sgn}(P)$. Changing the order of the factors in the wedge product we can extract $\binom{n}{r}$ independent products $dx^{i_1} \wedge dx^{i_2} \wedge \cdots \wedge dx^{i_r}$ with $1 \leq i_1 < i_2 < \cdots < i_r \leq n$. To each of these products corresponds in (6.28) the sum of $r!$ identical expressions

$$\frac{1}{r!}(\tau_{i_1 i_2 \cdots i_r} - \tau_{i_2 i_1 \cdots i_r} + \cdots)dx^{i_1} \wedge dx^{i_2} \wedge \cdots \wedge dx^{i_r}. \qquad (6.32)$$

Henceforth [CBDMDB91] capital letters will be used to label the set of $\binom{n}{r}$ ordered r-forms $\{dx^{I_1} \wedge \cdots \wedge dx^{I_r}; 1 \leq I_1 < \cdots < I_r \leq n\}$. Once the factor $\frac{1}{r!}$ has been eliminated, an arbitrary r-form $\tau \in \Omega_p^r(M)$ can be written as

$$\tau = \tau_{I_1 \cdots I_r} dx^{I_1} \wedge \cdots \wedge dx^{I_r} \qquad (6.33)$$

where the sum over all $1 \leq I_1 < \cdots < I_r \leq n$ is understood. This shows that (6.31) spans $\Omega_p^r(M)$.

In second place, we show that the set $\{dx^{I_1} \wedge \cdots \wedge dx^{I_r}; 1 \leq I_1 < \cdots < I_r \leq n\}$ is linearly independent. To start with, the assignment of the value zero to an r-form $\tau = \tau_{I_1 \cdots I_r} dx^{I_1} \wedge \cdots \wedge dx^{I_r}$ is taken to mean that the result of evaluating it on *any* set of vector arguments $(v_1, \ldots, v_r)$ is zero [CP86]. Consequently, if we consider any particular set of natural numbers $K_1, \ldots, K_r$ such that $1 \leq K_1 < \cdots < K_r \leq n$, the assignment of the value zero to $\tau_{I_1 \cdots I_r} dx^{I_1} \wedge \cdots \wedge dx^{I_r}$ should produce

$$\tau_{I_1 \cdots I_r} dx^{I_1} \wedge \cdots \wedge dx^{I_r}\left(\frac{\partial}{\partial x^{K_1}}, \ldots, \frac{\partial}{\partial x^{K_r}}\right) = 0. \qquad (6.34)$$

However, the evaluation of $dx^{I_1} \wedge \cdots \wedge dx^{I_r}(\frac{\partial}{\partial x^{K_1}}, \ldots, \frac{\partial}{\partial x^{K_r}})$ produces a non-zero value for $I_1 = K_1, \ldots, I_r = K_r$. This is so because, recalling (6.27),

$$dx^{K_1} \otimes \cdots \otimes dx^{K_r} \left(\frac{\partial}{\partial x^{K_1}}, \ldots, \frac{\partial}{\partial x^{K_r}} \right) = 1. \tag{6.35}$$

It follows that $\tau = \tau_{I_1 \cdots I_r} dx^{I_1} \wedge \cdots \wedge dx^{I_r}$ can only be zero if all the $\tau_{I_1 \cdots I_r}$ are zero. This proves that the set of r-forms $dx^{I_1} \wedge \cdots \wedge dx^{I_r}$ is linearly independent.

So, (6.31) defines a basis of $\Omega_p^r(M)$ with dimension $\binom{n}{r}$ which is the number of distinct combinations of r natural numbers $I_1 < I_2 < \cdots < I_r$ chosen from a total of n

$$\dim \Omega_p^r(M) = \binom{n}{r} = \frac{n!}{(n-r)!r!}. \tag{6.36}$$

The equality $\binom{n}{r} = \binom{n}{n-r}$ implies

$$\dim \Omega_p^r(M) = \dim \Omega_p^{n-r}(M). \tag{6.37}$$

In particular, $\dim \Omega_p^0(M) = \dim \Omega_p^n(M) = 1$ and $\dim \Omega_p^1(M) = n$. It is also convenient to define $\Omega_p^0(M) = \mathbb{R}$. So, a real valued C^∞ function on M is regarded as a 0-form. As $\Omega_p^1(M) = T_p^* M$, covectors are called 1-forms.

As (6.31) is a basis of $\Omega_p^r(M)$, the coefficients $\tau_{I_1 \cdots I_r}$, with ordered indices, can be rightly called components in the space $\Omega_p^r(M)$ and are known as *strict components*.

Example. Let ω and τ be two 1-forms, $\omega = \omega_{i_1} dx^{i_1}$ and $\tau = \tau_{i_2} dx^{i_2}$, in $\mathbb{R}^3$. Find the strict components of $\omega \wedge \tau$ in the basis $\{dx^{I_1} \wedge dx^{I_2}; 1 \leq I_1 < I_2 \leq 3\}$.
First we write

$$\omega \wedge \tau = \omega_{i_1} \tau_{i_2} dx^{i_1} \wedge dx^{i_2}. \tag{6.38}$$

As $\{dx^{i_1} \wedge dx^{i_2}; i_1, i_2 = 1, 2, 3\}$ is not a basis in the space $\Omega^2(\mathbb{R}^3)$, we cannot call $\omega_{i_1} \tau_{i_2}$ a component. But then, by changing the order of some factors in the wedge product in (6.38) we can extract

$\binom{3}{2} = 3$ independent products $dx^{i_1} \wedge dx^{i_2}$ with $1 \le i_1 < i_2 \le 3$. This corresponds to

$$\omega \wedge \tau = (\omega_{I_1}\tau_{I_2} - \omega_{I_2}\tau_{I_1})dx^{I_1} \wedge dx^{I_2}. \tag{6.39}$$

Here, $\{dx^{I_1} \wedge dx^{I_2};\, 1 \le I_1 < I_2 \le 3\}$ is our proposed basis for $\Omega^2(\mathbb{R}^3)$ and we can write in detail

$$\omega \wedge \tau = (\omega_1\tau_2 - \omega_2\tau_1)dx^1 \wedge dx^2$$

$$+ (\omega_1\tau_3 - \omega_3\tau_1)dx^1 \wedge dx^3$$

$$+ (\omega_2\tau_3 - \omega_3\tau_2)dx^2 \wedge dx^3 . \ \blacksquare \tag{6.40}$$

Chapter 7

Orientation of a Manifold

Let us consider a system of coordinates $(y^1, \ldots, y^n)$ in $\mathbb{R}^n$. Let us write down

$$x^i = x^i(y^1, \ldots, y^n) \qquad (7.1)$$

supposed to be single-valued, continuous, differentiable functions of $y^1, \ldots, y^n$. The non-vanishing, at any point, of the Jacobian

$$J = \det \left(\frac{\partial x^i}{\partial y^j} \right) \qquad (7.2)$$

occurs in the problem of the reversibility of (7.1) in terms of an n-tuple of *independent* coordinates $(x^1, \ldots, x^n)$

$$y^j = y^j(x^1, \ldots, x^n) \,. \qquad (7.3)$$

Indeed, the functions $x^i = x^i(y^1, \ldots, y^n)$ are non-independent if a differentiable function $F(x^1, \ldots, x^n) = 0$ exists and is satisfied whatever the values of $y^1, \ldots, y^n$. If this happens, F is independent of $y^1, \ldots, y^n$ and it follows that $\frac{\partial F}{\partial y^1}, \ldots, \frac{\partial F}{\partial y^n}$ are all zero. Therefore,

$$\begin{cases} \dfrac{\partial F}{\partial x^1} \dfrac{\partial x^1}{\partial y^1} + \cdots + \dfrac{\partial F}{\partial x^n} \dfrac{\partial x^n}{\partial y^1} = 0 \\ \cdots\cdots\cdots\cdots\cdots\cdots\cdots\cdots \\ \dfrac{\partial F}{\partial x^1} \dfrac{\partial x^1}{\partial y^n} + \cdots + \dfrac{\partial F}{\partial x^n} \dfrac{\partial x^n}{\partial y^n} = 0 \,. \end{cases} \qquad (7.4)$$

So, if the Jacobian determinant

$$\begin{vmatrix} \dfrac{\partial x^1}{\partial y^1} & \cdots & \dfrac{\partial x^1}{\partial y^n} \\ \cdots\cdots\cdots\cdots\cdots \\ \dfrac{\partial x^n}{\partial y^1} & \cdots & \dfrac{\partial x^n}{\partial y^n} \end{vmatrix} \tag{7.5}$$

does not vanish, the only solution of (7.4) is the trivial solution $\frac{\partial F}{\partial x^1} = \cdots = \frac{\partial F}{\partial x^n} = 0$ and there can be no functional dependence $F(x^1, \ldots, x^n) = 0$ of the functions $x^i(y^1, \ldots, y^n)$ [Nic61].

Note. Recalling the example in section 3.2, the transformation between spherical polar coordinates and Cartesian coordinates, in Euclidean 3-space, was only admissible once the restrictions

$$r > 0$$
$$0 < \theta < \pi$$
$$0 \leq \varphi < 2\pi \tag{7.6}$$

have been imposed. This guarantees that the Jacobian $J = r^2 \sin\theta$ of the transformation (3.21) does not vanish. ∎

Considering an open subset U of $\mathbb{R}^n$, two coordinate systems with coordinates $\{x^i\}$ and $\{y^j\}$, determining the associated ordered frames $\{\frac{\partial}{\partial x^i}\}$ and $\{\frac{\partial}{\partial y^j}\}$ on U, are said to define the same orientation if the Jacobian determinant $J = \det\left(\frac{\partial x^i}{\partial y^j}\right)$ is positive at all points of the subset. An *orientation* on U is the equivalence class of these ordered frames.

In 3-dimensional Euclidean space we distinguish between *right-handed* and *left-handed* orthogonal sets of vectors [CP 86] and a transformation is orientation preserving if the Jacobian is positive. This requirement is satisfied, in particular, by the transformation (3.21) which relates suitably restricted spherical polar coordinates with Cartesian coordinates.

Let us now extend the concept of orientation to an n-dimensional manifold M. Given any two charts $(U_\alpha, \varphi_\alpha)$ and (U_β, φ_β) with

$U_\alpha \cap U_\beta \neq \emptyset$, local coordinates $\{x^i\}$ and $\{y^j\}$ will correspond to a point $p \in U_\alpha \cap U_\beta$. The tangent space T_pM is spanned by either the ordered basis $\{\frac{\partial}{\partial x^i}\}$ or $\{\frac{\partial}{\partial y^j}\}$. The two bases are inter-related by

$$\frac{\partial}{\partial y^j} = \frac{\partial x^i}{\partial y^j} \frac{\partial}{\partial x^i} \tag{7.7}$$

where $\frac{\partial x^i}{\partial y^j}$ depends on p. If the Jacobian determinant $J = \det\left(\frac{\partial x^i}{\partial y^j}\right)$ > 0 at all points $p \in U_\alpha \cap U_\beta$, $\{\frac{\partial}{\partial x^i}\}$ and $\{\frac{\partial}{\partial y^j}\}$ are said to define the same orientation on $U_\alpha \cap U_\beta$. On the other hand, if $J = \det\left(\frac{\partial x^i}{\partial y^j}\right) <$ 0, $\{\frac{\partial}{\partial x^i}\}$ and $\{\frac{\partial}{\partial y^j}\}$ define opposite orientations.

A manifold M is connected if it is "in one piece", i.e., if it cannot be written as $M = M_1 \cup M_2$ with $M_1 \cap M_2 = \emptyset$. Let M be a connected manifold covered by an atlas of charts $\{(U_\gamma, \varphi_\gamma)\}$. If in the intersection of *any* two charts, there exist coordinates $\{x^i\}$ and $\{y^j\}$ such that $J = \det\left(\frac{\partial x^i}{\partial y^j}\right) > 0$, M is said to be *orientable*. This makes it possible to choose a consistent orientation in each pair of intersecting charts, patched together to define an orientation throughout the manifold M. An atlas with these properties is called an *oriented atlas* and then all coordinate transformations have positive Jacobian.

If M is covered globally by a single chart, like $\mathbb{R}^n$, it must be orientable.

An example of a non-orientable manifold is the Möbius strip [Nak90, NS83]. As a matter of fact, if one moves a basis frame along a path around the Möbius strip, it returns to the starting point with opposite orientation. Any parallelizable manifold is orientable. On the other hand, the 2-sphere S^2 although not parallelizable is orientable.

Now we want to show that the choice of a nowhere vanishing, smooth, n-form ω corresponds to another way of specifying an orientation of an n-dimensional manifold M [Mar91].

An r-form defined over M is a smooth assignment $\omega \in \Omega_p^r(M)$ to each $p \in M$. Let p lie in the domain U_α of a chart $(U_\alpha, \varphi_\alpha)$ corresponding to a coordinate system $x^1, \ldots, x^n$. To a non-vanishing n-form $\omega \in \Omega_p^n(U_\alpha)$ corresponds just one strict component

$(\dim \Omega^n_p(U_\alpha) = \binom{n}{n} = 1)$

$$\omega = \lambda_\alpha(p)dx^1 \wedge \cdots \wedge dx^n \tag{7.8}$$

where λ_α is a smooth non-zero function which depends on the chart. An ω defined over M is said to be smooth if the components are C^∞ for all charts.

A strictly positive λ_α for all $p \in U_\alpha$ can always be obtained by assuming a convenient choice of the coordinates. For example if $\lambda_\alpha(p) < 0$, permuting two consecutive coordinates makes $\lambda_\alpha(p) > 0$. Let (U_β, φ_β) be any other chart, with coordinates $y^1, \ldots, y^n$, such that $U_\alpha \cap U_\beta \neq \emptyset$. We can write for $p \in U_\alpha \cap U_\beta$

$$\omega = \lambda_\alpha(p)dx^1 \wedge \cdots \wedge dx^n = \lambda_\beta(p)dy^1 \wedge \cdots \wedge dy^n \tag{7.9}$$

and adopt coordinates such that $\lambda_\alpha(p) > 0$ and $\lambda_\beta(p) > 0$. As the change in coordinates induces a cobasis change (3.22)

$$dx^i = \frac{\partial x^i}{\partial y^j}dy^j \tag{7.10}$$

we can write for $p \in U_\alpha \cap U_\beta$

$$
\begin{aligned}
dx^1 \wedge \cdots \wedge dx^n &= \frac{\partial x^1}{\partial y^{j_1}}dy^{j_1} \wedge \frac{\partial x^2}{\partial y^{j_2}}dy^{j_2} \wedge \cdots \wedge \frac{\partial x^n}{\partial y^{j_n}}dy^{j_n} \\
&= \frac{\partial x^1}{\partial y^{j_1}} \cdots \frac{\partial x^n}{\partial y^{j_n}} \, \varepsilon^{j_1 \cdots j_n}_{1 \cdots n} dy^1 \wedge \cdots \wedge dy^n \\
&= \det\left(\frac{\partial x^i}{\partial y^j}\right) dy^1 \wedge \cdots \wedge dy^n .
\end{aligned}
\tag{7.11}
$$

Therefore, for $p \in U_\alpha \cap U_\beta$ it follows that

$$
\begin{aligned}
\omega &= \lambda_\alpha(p)dx^1 \wedge \cdots \wedge dx^n \\
&= \lambda_\alpha(p) \det\left(\frac{\partial x^i}{\partial y^j}\right) dy^1 \wedge \cdots \wedge dy^n \\
&= \lambda_\beta(p)dy^1 \wedge \cdots \wedge dy^n
\end{aligned}
\tag{7.12}
$$

where

$$\lambda_\beta(p) = \lambda_\alpha(p) \det \left(\frac{\partial x^i}{\partial y^j} \right). \tag{7.13}$$

So, once we have chosen coordinates such that $\lambda_\alpha(p) > 0$ and $\lambda_\beta(p) > 0$, it follows that for any pair of overlapping charts there exist coordinates so that $J = \det \left(\frac{\partial x^i}{\partial y^j} \right) > 0$, with the two domains being coherently oriented. If M is orientable then it possesses an atlas of coherently oriented charts. Then, we can employ a nowhere vanishing, smooth, n-form ω on M, to define a consistent orientation throughout M. It can be said that the specification of the n-form $dx^1 \wedge \cdots \wedge dx^n$ expresses an orientation on M.

In fact, an orientation corresponds to an equivalence class of n-forms; two n-forms differing from one another by an everywhere positive factor, expressing the same orientation.

Once an orientation has been specified in terms of an orienting n-form ω, an ordered frame $\frac{\partial}{\partial x^1} \cdots \frac{\partial}{\partial x^n}$, associated with a chart $(U_\alpha, \varphi_\alpha)$, is said to be positively or negatively oriented according to whether $\omega_p \left(\frac{\partial}{\partial x^1} |_p, \ldots, \frac{\partial}{\partial x^n} |_p \right)$ is positive or negative for all $p \in U_\alpha$. If M is an orientable manifold, there exists a selection of coherently oriented tangent spaces $T_p M$, for all $p \in M$.

Chapter 8

Hodge Star Operator

Let us consider the space of antisymmetric covariant tensors over a flat affine space modelled on an oriented n-dimensional vector space V endowed with a flat metric g, i.e., diagonal with the diagonal elements taking on the values $+1$ or -1.

Since

$$\dim \Omega^r(V,g) = \binom{n}{r} = \binom{n}{n-r} = \dim \Omega^{n-r}(V,g) \qquad (8.1)$$

the vector space $\Omega^r(V,g)$ is isomorphic to $\Omega^{n-r}(V,g)$.

Given an orientation and a metric we can set up an isomorphism

$$* : \quad \Omega^r(V,g) \to \Omega^{n-r}(V,g) \qquad (8.2)$$

which is called the *Hodge star operator* [GS87].

Let $\{dx^i\}$ be an orthonormal basis (4.56) of $V^* = \Omega^1(V,g)$ dual of V. Then we can build an orienting n-form $dx^1 \wedge dx^2 \wedge \cdots \wedge dx^n$ as mentioned in Chapter 7. The Hodge star operator can be defined by specifying the action on the basis elements $dx^{I_1} \wedge \cdots \wedge dx^{I_r}$ ($1 \le I_1 < \cdots < I_r \le n$) of $\Omega^r(V,g)$ as a $(n-r)$-form

$$*(dx^{I_1} \wedge \cdots \wedge dx^{I_r}) := g^{I_1 I_1} \cdots g^{I_r I_r} \varepsilon^{I_1 \cdots I_n}_{1 \cdots n} dx^{I_{r+1}} \wedge \cdots \wedge dx^{I_n} \qquad (8.3)$$

where $(I_1, \ldots, I_r)$ and $(I_{r+1}, \ldots, I_n)$ are sets complementary to each other in the set $(1, 2, \ldots, n)$. In Eq. (8.3) it is assumed that

$$1 \le I_1 < \cdots < I_r \le n \quad \text{and} \quad 1 \le I_{r+1} < \cdots < I_n \le n. \qquad (8.4)$$

Note that the definition of the Levi-Civita antisymmetric symbol $\varepsilon_{1\cdots n}^{I_1\cdots I_n}$ is tied up to the specification of an orienting form $dx^1 \wedge \cdots \wedge dx^n$ (Chapter 7).

It can be shown [BG80,GS87] that the definition of the operation $*$ is independent of the choice of the oriented orthonormal basis and Eq. (8.2) defines the Hodge star operator as a canonical isomorphism.

Example. In 3-dimensional Euclidean space with Euclidean metric $\delta = \text{diag}\,(1,1,1)$, *Cartesian coordinates* (x^1, x^2, x^3) and orienting form $dx^1 \wedge dx^2 \wedge dx^3$, the Hodge star of a 2-form is a 1-form

$$* : \Omega^2(\mathbb{R}^3, \delta) \to \Omega^1(\mathbb{R}^3, \delta). \tag{8.5}$$

For the basis $(dx^1 \wedge dx^2, dx^1 \wedge dx^3, dx^2 \wedge dx^3)$ on $\Omega^2(\mathbb{R}^3, \delta)$ we obtain

$$*(dx^1 \wedge dx^2) = \varepsilon_{123}^{123} dx^3 = dx^3$$

$$*(dx^1 \wedge dx^3) = \varepsilon_{123}^{132} dx^2 = -dx^2$$

$$*(dx^2 \wedge dx^3) = \varepsilon_{123}^{231} dx^1 = dx^1. \blacksquare \tag{8.6}$$

Example. Let us now apply to Minkowski space with metric $\eta = \text{diag}\,(1, -1, -1, -1)$, orthonormal coordinates $(x^0 = ct, x^1, x^2, x^3)$ and orientation expressed by $dx^0 \wedge dx^1 \wedge dx^2 \wedge dx^3$. Considering

$$* : \Omega^1(\mathbb{R}^4, \eta) \to \Omega^3(\mathbb{R}^4, \eta) \tag{8.7}$$

we readily calculate, for example,

$$*dx^0 = \eta^{00} \varepsilon_{0123}^{0123} dx^1 \wedge dx^2 \wedge dx^3$$

$$= dx^1 \wedge dx^2 \wedge dx^3$$

$$*dx^2 = \eta^{22} \varepsilon_{0123}^{2013} dx^0 \wedge dx^1 \wedge dx^3$$

$$= -dx^0 \wedge dx^1 \wedge dx^3$$

while taking

$$* : \Omega^2(\mathbb{R}^4, \eta) \to \Omega^2(\mathbb{R}^4, \eta) \tag{8.8}$$

we obtain, for example,

$$*(dx^1 \wedge dx^3) = \eta^{11}\eta^{33}\varepsilon^{1302}_{0123}dx^0 \wedge dx^2$$

$$= -dx^0 \wedge dx^2 \,. \blacksquare$$

Chapter 9

Wedge Product and Cross Product

In this chapter we will be working in the 3-dimensional Euclidean space which is a trivial example of differentiable manifold.

Cartesian coordinates determine an associated natural moving frame in $\mathbb{R}^3$. At each point p, the set of tangent vectors to the coordinate lines $\left(\frac{\partial}{\partial x^1}\big|_p, \frac{\partial}{\partial x^2}\big|_p, \frac{\partial}{\partial x^3}\big|_p\right)$, associated with the Cartesian coordinates, form a natural basis of $T_p\mathbb{R}^3$, orthogonal and with norm 1, in the usual correspondence with $(\hat{i}, \hat{j}, \hat{k})$. Let us now consider the operation which inverts all the coordinate axes, transforming a right-handed coordinate system into a left-handed one. Under this operation, known as inversion, *polar vectors* remain unchanged with their components changing sign at the same time as the axes are inverted. Polar vectors are therefore independent of the handedness of the coordinate system. However, under a change of handedness not all vectors transform in this manner. Let us consider, for example, the *cross product* of two vectors $u = u^i \frac{\partial}{\partial x^i}$ and $v = v^j \frac{\partial}{\partial x^j}$ in Cartesian coordinates of the 3-dimensional Euclidean space

$$
\begin{aligned}
u \times v = (u^2 v^3 - u^3 v^2)\frac{\partial}{\partial x^1} \\
+ (u^3 v^1 - u^1 v^3)\frac{\partial}{\partial x^2} \\
+ (u^1 v^2 - u^2 v^1)\frac{\partial}{\partial x^3}.
\end{aligned} \tag{9.1}
$$

These components make sense when compared with the components of the action of the Hodge star operator on the 2-form $u^* \wedge v^*$ (9.6).

65

Note. By a convenient choice of the Cartesian basis in 3-dimensional Euclidean space, we easily find the magnitude of the cross product as the product of the lengths of the two vectors by $\sin\theta$. ■

The components of the vector (9.1) do not change sign under inversion. Vectors with this property are called *axial vectors or pseudovectors*. Thus, if we change the handedness of the coordinate system, axial vectors change into the opposite vectors. That is, the "arrow" can only be specified once we specify the handedness of the coordinate system. They are tied up to a twisting choice.

In the particular case of the 3-dimensional Euclidean space, the wedge product of two 1-forms, compounded with the Hodge star operation, yields a 1-form whose associated vector reproduces the properties of the familiar cross product of the 3-dimensional Euclidean vector algebra [CP86, Mar91, Sch80]. Since the Hodge star operation is defined with respect to a given orientation, its effect on a 2-form is to produce a 1-form associated to an axial vector in the assumed orientation. Besides, it is only for $n = 3$ that $\Omega_p^2(\mathbb{R}^n)$ and $\Omega_p^1(\mathbb{R}^n)$ have the same dimension, $\binom{n}{1} = \binom{n}{2}$, making it possible to establish, by means of the Hodge star operator, the canonical isomorphism

$$* : \Omega_p^2(\mathbb{R}^3) \to \Omega_p^1(\mathbb{R}^3) = T_p^*\mathbb{R}^3. \tag{9.2}$$

Furthermore, as seen in Chapter 4, by the metric dual operation which correlates, in an unique way, elements of $T_p\mathbb{R}^3$ and $T_p^*\mathbb{R}^3 = \Omega_p^1(\mathbb{R}^3)$, we can associate to the vectors $u = u^i\frac{\partial}{\partial x^i}$ and $v = v^j\frac{\partial}{\partial x^j}$ the 1-forms $u^* = u_i dx^i$ and $v^* = v_j dx^j$. Considering the wedge product of the two 1-forms we obtain a 2-form in $\Omega_p^2(\mathbb{R}^3)$ that can be expressed in terms of the strict components (Eq. (6.40)) in the basis $(dx^2 \wedge dx^3, dx^1 \wedge dx^3, dx^1 \wedge dx^2)$ as

$$u^* \wedge v^* = (u_2 v_3 - u_3 v_2) dx^2 \wedge dx^3$$
$$+ (u_1 v_3 - u_3 v_1) dx^1 \wedge dx^3$$
$$+ (u_1 v_2 - u_2 v_1) dx^1 \wedge dx^2. \tag{9.3}$$

But, since we are working in Cartesian coordinates (x^1, x^2, x^3), the matrix (g_{ij}) is regarded as the identity matrix and we can write as in Chapter 4

$$u_i = u^i \quad \text{and} \quad v_j = v^j. \tag{9.4}$$

As a result, by acting on the wedge product $u^* \wedge v^*$ with the Hodge star operator

$$* : \Omega_p^2(\mathbb{R}^3) \to \Omega_p^1(\mathbb{R}^3) \tag{9.5}$$

we obtain by Eqs. (8.6)

$$*(u^* \wedge v^*) = (u^2 v^3 - u^3 v^2) dx^1$$
$$+ (u^3 v^1 - u^1 v^3) dx^2$$
$$+ (u^1 v^2 - u^2 v^1) dx^3. \tag{9.6}$$

This is a 1-form whose components are equal to the components of the cross product $u \times v$ (Eq. (9.1)) in the basis $\left(\frac{\partial}{\partial x^1}, \frac{\partial}{\partial x^2}, \frac{\partial}{\partial x^3} \right)$.

Summing up, it is only when $n = 3$ and once specified an orientation, that $*(u^* \wedge v^*)$ defines a 1-form which is the metric dual of the cross product $u \times v$ as mentioned in (9.1).

$$*(u^* \wedge v^*) = (u \times v)^*. \tag{9.7}$$

The object of this exercise is to present a comprehensive review of the geometrical properties of vectors and covectors (1-forms). The relation between the wedge product of two 1-forms and the cross product of the associated vectors, in 3-dimensional Euclidean vector algebra, emphasizes the interpretation of the wedge product as a more general concept than that of cross product. It becomes clear that the cross product corresponds to an axial vector and why it is not generalizable to spaces other than the 3-dimensional Euclidean space.

Bibliography

Abraham R. and Marsden J. E. (1978). *Foundations of Mechanics*, Second Edition. Addison-Wesley, Reading, Massachusetts.

Abraham R., Marsden J. E. and Ratiu T. (1988). *Manifolds, Tensor Analysis and Applications*. Springer-Verlag, New York.

Aldrovandi R. and Pereira J. G. (1995). *An Introduction to Geometrical Physics*. World Scientific, Singapore.

Benn I. M. and Tucker R. W. (1987). *An Introduction to Spinors and Geometry with Applications in Physiscs*. Adam Hilger, Bristol.

Bishop R. L. and Goldberg S. I. (1980). *Tensor Analysis on Manifolds*. Dover Publications, New York.

Boothby W. M. (1986). *An Introduction to Differentiable Manifolds and Riemannian Geometry*. Academic Press, New York.

Choquet-Bruhat Y., DeWitt-Morette C., and Dillard-Bleick M. (1991). *Analysis, Manifolds and Physics*. North-Holland, Amsterdam.

Courant R. (1959). *Differential and Integral Calculus*. Blackie & Son Limited, London.

Crampin M. and Pirani F. A. E. (1986). *Applicable Differential Geometry*. Cambridge University Press, Cambridge.

Curtis W. D. and Miller F. R. (1985). *Differential Manifolds and Theoretical Physics*. Academic Press, New York.

Frankel T. (2004). *The Geometry of Physics*. Cambridge University Press, Cambridge.

Göckeler M. and Schücker T. (1987). *Differential Geometry, Gauge Theories and Gravity*. Cambridge University Press, Cambridge.

Isham C. J. (1989). *Modern Differential Geometry for Physicists*. World Scientific, Singapore.

Mansfield M. J. (1963). *Introduction to Topology.* D. Van Nostrand Company, Princeton, New Jersey.

Martin D. (1991). *Manifold Theory.* Ellis Horwood, New York.

Misner C. W., Thorne K. S., and Wheeler J. A. (1973). *Gravitation.* W. H. Freeman, San Francisco.

Munkres J. R. (2000). *Topology.* Prentice Hall, New Jersey.

Nakahara M. (1990). *Geometry, Topology and Physics.* Adam Hilger, New York.

Nash C. and Sen S. (1983). *Topology and Geometry for Physicists.* Academic Press, New York.

Nicolson M. M. (1961). *Fundamentals & Techniques of Mathematics for Scientists.* Longmans, Green and Co. Ltd, London.

Schutz B. F. (1980). *Geometrical Methods of Mathematical Physics.* Cambridge University Press, Cambridge.

Sokolnikoff I. S. (1964). *Tensor Analysis.* John Wiley & Sons, New York.

von Westenholz C. (1981). *Differential Forms in Mathematical Physics.* North-Holland, Amsterdam.

Index